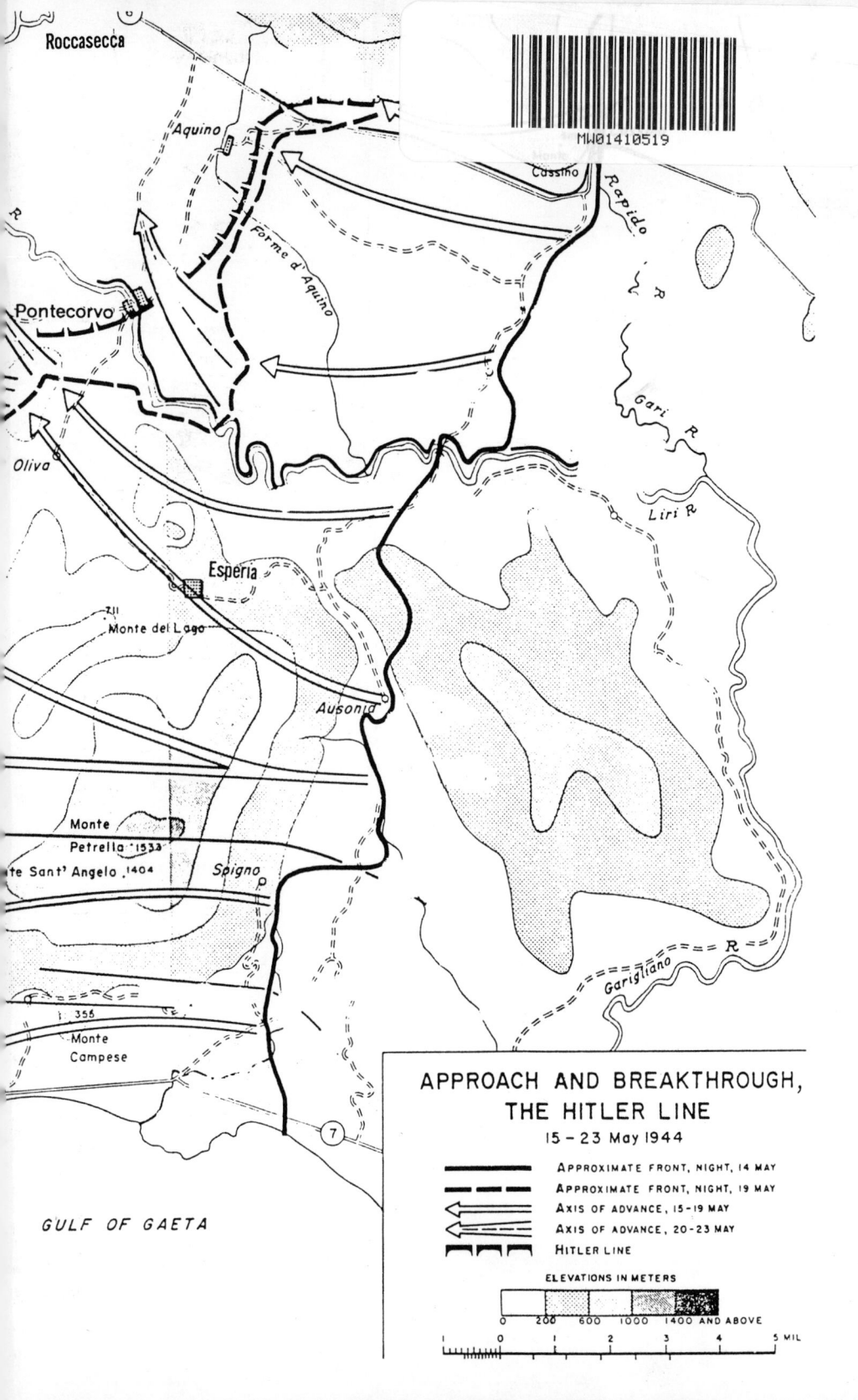

PICO

PICO

The White Paper Act

John A. Abatecola

San Diego, CA

"Pico"

Copyright © 1998 by: John A. Abatecola

Tadone Publishing
2707 Congress St.
San Diego, CA 92110

All rights reserved

No part of this book may be duplicated or utilized in any manner, except for brief quotations for reviews & critical articles. The author also welcomes any comments or suggestions.

Library of Congress: Cataloging - in - Publication Data
ISBN 09659628-0-6
Produced by Windsor Associates, San Diego, CA

First Edition

Joann Sanchez, Editor

Jacket Illustration by: Lisa and Diana Abatecola

Printed in the United States of America

Dedicated to my beloved Elena,
for her courage, understanding and love.

PREFACE

This book is based on a true story. Most of the events occurred during the period 1939 to 1948.

Throughout the history of wars, it is a known fact that conditions, in the form of plunder, foraging, starvation, rape etc., are prevalent with the so-called "territory" of a conquered nation.

If indeed one can describe a normal conflict, it would constitute a battle between soldier of opposing parties.

Modern war consists of Air, Naval and Ground Power that do not intentionally injure or kill civilians.

However, mistakes do occur and non-combatants become casualties of unavoidable events. There are times when invading armies are allowed to plunder the civilian population of the conquered enemy.

Situations have happened where an advancing army is stopped because the enemy has prepared its defenses well. In the case of the battle of Monte Cassino, the Germans established and strategically fortified the high ground throughout the entire mountainous area. They pre-engineered portable bunkers that were placed into dug out areas of the mountain sides. They expanded rock caves that were also utilized as living quarters. Their observation posts were equally fortified. In actuality, their placements were impregnable. To make matters even worse for the Allies, the Germans were excellent disciplined soldiers who adapted very well to their mountainous surroundings. They had one slight disadvantage compared to the Allies, they were low on ammunition.

The Allies had nearly everything against them. They had the low ground which in this case had the Allies under constant observation. The weather was always cold, damp and muddy. They had to dig-in where it was nearly impossible because of the rocky terrain. Their supply lines were critically hampered because the enemy annihilated most of the mules with mortar and artillery fire. The Allies had the only air power in the theater of action but it was rendered useless because of the close proximity between adversaries. Estimates are that the Allies lost nearly as many casualties to the hostile elements as compared to the enemy.

Chances were that it would take a force of five Allies to dislodge one deeply-entrenched German soldier. Compared to the Allies, the enemy was certainly enjoying conditions of luxury.

The entire defensive position was called The Gustav Line by the Germans. The task facing the Allied Armies was to break the line and quickly advance to the Liri Valley and, consequently, Rome.

The Allies were doing very well on all fronts throughout the war. The only snag was the campaign in Italy namely, Cassino. The Allied Commanders in the area were under considerable pressure to break the stalemate. There were rumors that high echelon command replacements were a possibility. An assessment of the overall operation was that they were not winning. In fact, if the casualty ratio were utilized as a barometer, the Allies were losing.

The Allied commanders were debating on whether or not to bomb the ancient Monastery located on the top of Monte Cassino. There were both pros and cons involved in this decision.

The Allied command insisted the Germans were utilizing the Abbey as an observation post. Others believed that enemy occupied it as a fortified military position. Some insisted that the Germans were nowhere near the walls of the Abbey.

The exposed Allied ground soldiers blamed and cursed the monastery every time an artillery round landed near them. The Germans swear that they never used the area in question as an observation post. Needless to say, the debate continued as the pressure increased on the commanders.

The casualties mounted to an alarming level. The ironic factor was that the distraction had no significant value or result in breaking through the Gustav Line. The enemy had enough observation posts and consequently did not need the Abbey to create precision havoc over every square yard of Allied-held position. They destroyed the Abbey with saturation bombing with the most tonnage of bombs ever dropped on one building! After completing the destruction, the Allies failed to take advantage of the bombing. The normal tactical procedure is to attack the enemy after he has been reeling from heavy bombing. The method has always been an advantage to the attacking force. They never attacked Monte Cassino. Either a mistake or by choice, the Allies attacked a nearby hill and were repelled with enormous casualties.

Were the Allied commanders incompetent or were they laying down an efficient smoke screen that commanded the headlines?

The Monastery on Monte Cassino was famous throughout the Christian World. It made the headlines in newspapers. The question of bombing had political ramifications. The entire building was a museum and subsequently, a great contribution to western civilization. Fortunately, the Germans had removed all the movable art and priceless manuscripts for safekeeping.

Was it all a smoke screen to shield the cover-up for the drastic events that were in the planning stage? The enemy was solidified in their positions that the stalemate could continue indefinitely or at least, to the end of the European war.

Marshall Juin was the Commander of the French Expeditionary Corps. His evaluation of success was based upon the brutality and rapidity of the attack! For months, he had been advocating a flanking move in the most difficult of the mountainous areas. His military strength consisted of Moroccan Goumiers. There is no question that they were the best mountain fighters of the entire campaign. They adapted themselves to exist for days on small amounts of food and water. They traveled light and fast but most of all, they were indeed brutal and possessed the lowest moral values of anyone associated with war in the European Theater. They were the right force in an adaptable environment versus a formidable foe but not the correct civilian population. The French Expeditionary Corps plan was excellent, except that it lacked an incentive for the brutally-inclined Moroccan Goumiers.

Somewhere along the planning stage, an incentive was created and subsequently implemented into the breakthrough operation. The Italian civilian population refer to it as, "The White Paper Act." It explained that as long as these Goumiers and associates were advancing for a period of seven days, they could unleash their brutality in the form of rape and other atrocities. It is not clear who authorized the White Paper Act, but it happened in all of the little towns that were in the path of the Moroccan Goumiers. The plan was a complete success. It was the primary factor in the German defeat in Southern Italy. Needless to say, the victory was at the expense of a friendly Allied population.

The brutal atrocities occurred in the form of rape to the worst degree. They raped children, young women, old women, boys, men and even animals! They raped an eighty year old woman a total of six times. Their lack of respect for religion didn't stop them; they raped nuns.

The Italians were now part of the Allied fighting force. Over 300,000 Italian soldiers were armed by the Allies. In fact, there was at least one Italian Division facing the Germans on the Gustav

of trouble throughout Italy. The populations of these mountain hamlets were anticipating freedom from the Allies. Instead, they received the stigma of lingering nightmares that will accompany them for a lifetime. The atrocities came to pass as an event but not as a memory. The theater commanders subsequently conquered Rome.

This is the story of Pico, Italy. It is a story that must be revealed to the most humane of all nations: the United States.

ACKNOWLEDGEMENTS

I wish to thank the following individuals. Some are close friends who encouraged me to write this book. Others whom I am grateful to include the many new acquaintances that gave me the inspiration to complete the work. I was exceptionally fortunate to have the encouragement and support of my immediate family of Elena, Anthony, Lisa, Diana, Eleisa, Lauren and Jon-Michael. Further gratitude to my brothers, Michael Abatecola and William Abatecola. Most of all, to my exceptional mother Natalie; thank you for being my mother.

Sumner "Charlie" Brown, Biondi & Mary Pizzi, Albert N. & Anna Leschi, Phillip Germani, Carl Plett, Robert Munroe, Janet Jenkins, Anthony Gelsomino, Spencer Joseph Speer, Luigi Luevano, Gino Greco, Tulio Cerceillo, Glen Spangler, Karen Spring, Barbara Holt, Debrah Blackwell, Antonio Monti, Amy Gardner, Mike Castelucci, J.W. August, Jerry G. Bishop, Chet Forte, Jerry Durnell, Larry Denmark, Michele Leschi, Peter Tabor, Robert "Pop" Veyera, Jeffery Morton, John Moriarty, P.N. Sukumar, Thomas J. Nemeth, Linda K. Sanborn, Rose A. Favale, Sam Marasco Sr., Joseph Ciokin, Rinaldo Abatecola, Peter Tabor, Bobby "Pop" Veyera, John Rendine, Henry Pezza, Horst Hindinger, Albert Banna, John Ciangiarulo, Frank Bandinelli, Henry Tarlian, Harry Kant, Junior Biery, Frank Roselli, Nick Milakovich, John Quarnstrom, Sandy Nelson, Hilton Vail, Jack Borroff, Mike Cahill, Matt Fried, Len Hendren, Jerry Miller, Marvin Mears, Nels Normann, Dennis Sciotto, Fred Walters, Nick Zizzo, Bruce Zissen, Allison Barksdale, Steve Rouse, Ralph Falacana, Jeff Morton, John McPeat, Derek Gerber, Babe Micelli, Marianne Cadiz, Barbara Holden, Reta Atkisson, Michael Munoz, Sam Antonio, John Pellicano, Carl Archelitti, Fred Walters, Mike & Geneiveve Matherly, David McGee, Sally Lombardo, Susan A. Sherman, Rene Conti, John Andolina, Richard Conti, Krystine Towne, Karen Conti, Loretta Buongiorno, Louis Carnevale, Scott Cino, Anna & Jay Hammond, Doug & Star Kerr, Peter Cino, Iona Sampson, Pauline Bovin, Richard & Pauline Marso, John Conti, Tom & Maryln Iacovetti, Angelina & Roy Palmiciano, Santina Forti, Ronald Stout, Nick Zizzo, my cousins residing in the United States, my cousins living in Italy and above all, the Peasants I had the pleasure of meeting during my recent trip to Italy. Most of the story is based on their lives. They did not disappoint me in the least. I would be proud and honored to live among them in my beloved PICO. I apologize for names I've forgotten to mention, but the old adage, "Out of Sight, Out of Mind" commands one's memory.

1
Justifiable Vengeance

A large Italian transport ship was approaching the Statue of Liberty in New York harbor. It was the spring of 1931. The ocean voyage had been an uncomfortable experience because of the stormy conditions. The accommodations were poor for the third class passengers but the trip was finally over.

The pregnant Anna Micheletti held the hand of her six-year old niece, Gabriella. The little girl's mother, Anna's sister died at childbirth. Her father was killed in battle during World War I.

Anna raised her sister's child since birth. Gabriella was her daughter.

"Hurry or you will miss seeing The Lady," ordered Anna as she squeezed the little girl with happiness.

The day was overcast but to them the sun was shinning brightly. She turned to Gabriella and said, "This is America. You will have many opportunities here. The only thing missing is your mother and father. They would have loved this moment."

"Mama, look, the buildings are so high," the child clapped her little hands and announced as if she understood, "I'm so happy to be in America."

Anna beamed with elation, "This land is sometimes called God's Country. Everything that is good happens here."

Gabriella asked, "Why does God love America more than all the other places?"

Her mother looked into her quizzing eyes, "I have never thought about what you ask but I think I can give you a good answer."

Pico

"What, Mama?"

"People from all over the world come here to live. They have different ways about themselves and speak different languages but are accepted here. God is so happy that all his people can live with each other that He has given a special blessing to this land."

The child replied, "God is good."

Anna made the sign of the cross as an approval gesture to Gabriella's appraisal of the Almighty. A tug came alongside and the harbor pilot came aboard to steer the ship dockside.

Her husband, Angelo, resided in America for the past eleven years. He occasionally went back to Italy as an obligation to keep his wife pregnant.

Five years previously, he had sent for their two sons and a daughter who lived with their father in New England.

Anna had some doubts about Angelo. He was a handsome man who had a reputation as a womanizer. She intended to change his lifestyle but realized that any change for Angelo would be a challenge. Anna was well aware of the mind-set associated with the average adult Italian male. Their concept of marriage was, at best, confusing. In most cases, it was a one-way proposition favoring the male partner. He thought nothing of having affairs with other women. Of course, he was discrete about it but if his wife discovered his extracurricular activities, she was expected to look the other way. His explanation was that his wife was on a pedestal and he loved her above all. However, if the situation was reversed, all hell broke loose. An affair by an Italian wife was out of the question. Indeed, if such an event occurred, she could become an outcast even to her best friends. Needless to say, the male was praised and the woman doing the same thing was scorned. Anna missed her children and looked forward to the happy occasion.

Angelo was having trouble getting her to immigrate because she wouldn't leave her old parents. Six months ago, he decided to rectify the problem. He went back to Italy determined to convince his wife to come to America. Her parents insisted that she comply with Angelo's wishes. They all agreed that Anna's property should be retained until she was sure that the move was permanent. The property next door to Anna's was owned and occupied by her parents.

Justifiable Vengeance

While Angelo visited Italy, Anna became pregnant, a standard formality. Gabriella, behaved and watched the statue. "The baby inside me is going to be an American and I know that either he or she is absorbing the scenery."

"Mama, do you really think so?"

Anna hugged the little girl and smiled.

Unknown to Anna, a young admirer from her village circulated lies through a cousin in America that he and Anna were lovers. The lies reached the ears of Angelo during the ocean voyage. She was true to her husband. He was the only man she ever had. The naive Angelo believed the false rumors. When the ship arrived, he was waiting in anger. As she came down the gangplank, she noticed that her children were missing. Anna hesitated a moment and then ran to her husband with open arms. Angelo had no intentions of questioning her of the rumors. He was embarassed by the alleged rumor and consequently made his decision.

He eluded her embrace and slapped her hard and blood surfaced from the side of her mouth. She was startled as he stepped back and looked her in the eye, "Brutta putana."

Gabriella began to cry as Anna asked, "What is wrong with you? Have you lost your senses?"

People had gathered and were drawn to the commotion as his wrath continued, "Avanti. Here are your return tickets. I don't want a putana who has been sleeping with Tino Baldini."

Anna began to cry, "I have never slept with anyone."

He motioned to slap her again but chose words instead, "Putana, avanti. I don't ever want to see you again, even if I'm on my death bed."

"What about the children?"

"Putanas don't have children, they only have a good time."

"Don't I have a chance to answer your accusations?"

"America is too good for you. I will not allow the children to grow up knowing that their mother is a whore."

Anna realized that he was deranged but asked, "What about the child I have in my womb?"

He replied arrogantly, "It belongs to you and whoever is responsible for it." Anna screamed a loud piercing shrill and spit into his face. Gabriella cried as Anna took her hand, turned and walked away .

Pico

She had an entire voyage back to think and originate a plan of action. Anna's plans were to confirm the rumors originated from Tino Baldini. He had tried many times to initiate an affair with her but she never considered it. Tino would rush to the well in the village square to help her but she managed to evade his advances. He once caressed her leg and she responded with a slap across his surprised face.

That day, he explained in anger, "Angelo is sleeping with most of the women in America and you don't do anything about it."

She replied, "Go to hell."

As she sat dejectedly on deck trying to evaluate her recent experiences, a well-dressed gentleman paced back and forth, displaying a smile.

Anna returned a cold stare and ignored him. She thought: Why did Angelo pull such a ridiculous stunt? Could he be over-macho and stupid or does he have a steady woman in America? The asino must be stupid because I think I'm superior to anyone he knows here.

The gentleman walked by again and tipped his hat. Anna sneered at him. She took a deep breathe of the invigorating air and rationalized: Everything happens for the best. I never wanted to leave Italy in the first place.

When they returned to the village, she used the excuse that her papers were not in order and consequently were not allowed to debark. After a few weeks, the true story came out but no one ever confronted Anna about it. They all knew that she was true to Angelo and considered him an idiot.

It wasn't long before the baby was due to be born. Antonio was looking foward to the great event. He had a strong feeling that this grandchild would never leave Italy. Besides Anna and the newborn would be closest to his heart.

It was a clear, quiet night. Anna had labor pains throughout the day and was being attended by a midwife.

Antonio had summoned his friend who was an astrologer. As Antonio was smoking his pipe, the astrologer was observing the celestial heavens which displayed thousands of visible lights.

The women were indoors as candlelight flickered faintly through the windows. They all waited for the moment as the midwife walked back and forth. She approached Anna and inspected her. "It's not time yet, but it's getting close."

Justifiable Vengeance

The peasants put great faith in the astrologer's interpretation regarding a new birth. The exact moment determined the future years. The stars had to be located in a precise and favorable position at the time of the birth.

Antonio re-lit his pipe and asked his friend, "Are the conditions right or are we in trouble?"

"Antonio, I'll be honest with you. We have approximately four minutes. The closer we are to the fourth minute, the more perfect the future."

Antonio was ready to panic, "What happens if we exceed four minutes?"

"Everything will be bad unless she can hold it for tomorrow night."

The astrologer rose to his feet and pierced his eyes in silence. He gave a quick glance at his watch and then back to the star studded sky.

The silence was broken with voices from the house as the old men looked at each other. Immediately, the sweetest of baby voices penetrated the still night. Antonio looked at the astrologer awaiting the anticipated interpretation.

The astrologer's eyes left the heavens and beamed with joy, "This is the most unbelievable event that I've ever experienced."

"Is it good news?"

"Antonio, don't you understand, it was perfect! The stars were aligned right on target. It was impossible to be better. I have never heard of the perfect time, let alone being the astrologer, for a miracle."

Antonio asked happily, "What does it mean?"

"The child is special, an angel directly from heaven."

One of the old women came out and announced, "It's a girl and she is healthy."

Antonio had a bottle of prized wine nearby that he held until the happy event was over. He poured two glasses and congratulated the astrologer. The stargazer touched glasses and said, "Here is to a newborn angel."

"Yes," replied Antonio as he raised his glass, "to baby Angelina."

"Tell me, what lies in the future for my Angelina?"

"Everything good. She will be intelligent, beautiful and loved by all. She will be strong, highly respected and famous. Your Angelina will marry well and be happy with many children."

"I want to believe you," Antonio said.

Pico

"You must believe me because I have never been more convinced than the results of this memorable evening." The old men emptied two more bottles of wine as they celebrated into the night. Anna was up and around the following day. She was proud of her Angelina. When her father explained the astrologer's interpretations, she readily agreed to the name. During the following weeks, she made friends with Tino. Anna was determined to lure him into a pre-planned trap. She smiled and joked with him. One day he asked her, "When are we going to share a bed?"

She quickly responded, "I just gave birth, can't you wait until I heal?"

"I can wait, so take your time," he responded.

He walked away in anticipation of sexual victory. Anna watched him disappear down the trail and spoke under her breath, "You are going to be one sorry son of a bitch."

Late summer in Pico is the occasion for the Feast of San Antonio. In the eyes of many, one of the most joyous periods of the year. An event happened on the eve of the religious festival that would endure the test of time until replaced by a more memorable occurrence.

The evening was humid from the dampness that rolled off the Aurunci mountains. In fact, the sound carried well throughout the hills. On this night, one could hear the faint sound of the distant train.

Anna was now sharing a bed for the first time, with her soon to be ill-fated lover, Tino. The fog had passed over their area and the moon displayed a clear full brightness. Tino kissed her and said, "I've always wanted you and only you."

She replied with a forced giggle, "I'm a middle-aged woman with four children."

"Ah, but you are the most beautiful woman in all of Italy."

"Tino, you are so kind and so loving. Turn over and I will massage you as you fall asleep."

The young handsome man complied as Anna ran her fingers down his massive shoulders. He asked, "Why didn't we begin this wonderful relationship sooner?"

"Because I was never aware of you until I came back from America."

"Didn't you like America?"

"No, now relax," as she spoke with soothing and assuring words. "Sleep, my love, and before the sun rises, I will have a big surprise for you."

Justifiable Vengeance

Tino rolled on his back as the moonlight glowed through the open window. She lay next to this man with vengeance in her heart. Anna had no intentions of sleeping. Tino had unknowingly destroyed her family. He had forgotten about the lies that he circulated months ago. Anna had tears in her eyes as she looked at Tino sleeping next to her. She lay on her side, her beautiful body gracefully outlined by the penetrating moonlight. Anna thought to herself and recalled the events that occurred with the fiasco to America.

There was no question regarding the fate of Tino Baldini. Anna made up her mind during the long voyage home: Why kill him? He needs to face future life bearing a cross confirming his dishonorable actions. She had no remorse in what was about to happen. Anna was calm and confident. If the facts resulting in what she was about to do were made public, Anna knew that the people would forgive her.

She moved quietly out of bed. Tino stirred and asked sleepingly, "Where are you going, my love?"

"I need to use the bed pan."

Tino rolled over on his back as Anna dressed unnoticed, in her underclothes. She removed the sharp straight razor from her handbag. She recalled the many times she saw her father shaving with the same instrument. La Signora Micheletti slipped back into bed and snuggled with Tino. He moaned with pleasure and was in a state of euphoria. Tino had the very person that he desired. She decided the time was right. It was an hour before dawn. The frogs and crickets were challenging the nightingales in disrupting the eerie state of serenity. Anna displayed deep confidence as she initiated a brief career as a surgeon. She massaged Tino's bare stomach as he replied with a soft moan of pleasure. Anna took careful hold of his manhood and with a swift precision cut, accomplished her mission.

For a split-second, there was no reaction from Tino. She threw his manhood into his face. Tino exploded with a scream that awakened the entire town. The screams pierced the stillness of the night and carried resoundingly into the nearby mountains.

Anna walked boldly to the door and spoke over his deafening cries, "Now, no one can ever believe you if you ever again circulate lies."

She calmly left and walked unassumingly through the awakening village.

Pico

Anna could hear the faint screaming as she arrived home. The eastern sky lighted over the Abruzzi Mountains. She quietly enjoyed the serenity. The screaming subsided. A new day dawned. It was the feast of San Antonio.

Anna would become the object of unwanted folklore. Her family as a unit was destroyed. Vengeance was the option that she accepted. Hopefully, time or future events would heal and subsequently prevail.

She was prepared to become stern and determined. Her husband, Angelo could never be forgiven. He prejudged an innocent caring wife and mother guilty. Anna vowed that she, Gabriella, and baby Angelina would never leave Italy. In future years, she blamed Angelo for nearly every problem she encountered.

Her train of thought was suddenly interrupted by the crying baby. Anna went to the little bed and cuddled the little girl in her arms. She smiled and tickled the baby under the chin, "I love you so much, my little angel."

The baby responded with a smile as Anna began breast-feeding her.

Gabriella had awakened rubbing her innocent eyes, "Where were you, Mama, I looked for you and you were gone?"

Anna took her into her left arm and assured her, "I will never leave you or Angelina alone ever again. I had some important business to settle, you wouldn't understand."

Gabriella hugged her tightly, "I love you, Mama."

"I love you, too."

She placed baby Angelina softly back to bed as her best friend Anita entered with a sly smile that she partially covered with her hand. Anna made some thick coffee. They both sat outdoors and enjoyed the coffee along with the spectacular eastern sunrise breaking over the ever present Abruzzi Mountains.

"It's a beautiful day for the Festival," Anita said.

"Yes, it is. It's my favorite time of the year."

"Are you going to attend and take part in the festivities?" Anita asked.

"Of course I am, why shouldn't I?"

"Anna, I'm here because I'm your best friend. The entire town is probably now aware of what happened to Tino. They don't know why or who but I figured it out."

Justifiable Vengeance

Anna laughed playfully, "Good for you, are you still my friend?"

She took Anna's hand. "If I were in your shoes, I probably would have acted the same way except that I don't have the nerve."

"Thank you Anita, I feel at ease having a true friend."

"You do know that people will talk behind your back."

"I don't care what people gossip about. If they desire, they can write a song about me."

Gabriella was listening and asked, "What did you do, Mama?"

"Go play with the baby. Little children should be seen and not heard."

Anita asked, "Do you think the polizia will pay you a visit?"

"Not a chance, don't be silly. Tino is very chauvinistic. That type is so embarrassed that he will leave town as soon as he's able. Tino will never reveal what happened."

"But Anna, if the word gets out, previous events will bear out that you are the suspect."

"I don't care. If anyone points a finger at me, I'll deny it. I don't feel the least bit bad for him. The asino destroyed my family. He had no right to do that. As for Angelo, he deserves the same fate."

"Anna, he is your husband. Think about your children in America."

"Talk about nerve, I was willing to forgive him for sleeping with women in America but not anymore. If my children there love me, they will return."

"Anna, I'm sorry to say it but once they taste America, they will never return. Perhaps, once he hears the truth, he will want you back."

"Never, we are finished. I don't ever want a man again. Sex is over for me. After what I've been through, I have no desire whatsoever."

Anita respond defiantly, "You are too beautiful, you'll change."

"No, I will dedicate my life to raising my two daughters. Between them and operating this farm, I'll be too busy to think about sex. I can take it or leave it." The beautiful and vivacious Anna decided on the latter.

In later years, Angelo wrote her letters for reconciliation. She opened the correspondence only to remove the accompanying money. Anna then proceeded to discard the letter into the fireplace. She never read them.

During one of the occasions Gabriella asked, "Mama, don't you want to hear about my brothers and sister?" Anna wiped the tears from her eyes as she embraced the inquisitive little girl and explained, "If they love their mama they will come back to visit."

Approximately six years passed and Anna was true to her dedication. She was rearing two fine daughters. The Tino Ballad as written was true to Italian custom. Anna was a heroine to the female populace. Her young daughters were admired by all.

THE BALLAD OF OUR AUNT

Zia lodato agni momento,

Masto Tino non cia niente,

Zia lodato sempre zia,

Glia tagliato nostra-Zia,

Zia lodato agni momento,

Masto Tino non cia niente,

Zia lodato sempre zia,

Glia tagliato nostra-zia.

2

Hero In The Making

The location is a southern New England town. Mario Calcagni was a young boy of fourteen. He resided in a section of town that consisted of an eighty percent Italian-Americans.

The depression years directed the life-style of this community as well as similar locations throughout the nation. The people were poor but they had a majority of company; nearly everyone was at poverty level. Employment was scarce for the multitudes.

However, in regards to the Italians, among the Portuguese, the Polish and especially the Colored; there was hardly any employment.

Mario once overheard the older men talking about standing in a W.P.A. line in anticipation of obtaining whatever work was available.

When an Italian reached the employment window and was asked his name he was discouraged and ordered to go back to the end of the line. The men had patches on their pants but they were not embarassed because their peers also had patches. The women wore the same dresses every day. They washed their clothing nightly and hung their garments on the clothesline. The average Italian family planted gardens and had a few chickens and a supply of eggs. It was important to raise a sizable crop for canning and storage for the winter. Very few Depression children ever tasted butter and most never heard of bacon. The Italian families of the area, however, had one distinct advantage over others. Pasta was inexpensive to make and they cultivated an abundance of tomatoes. Meat was prohibitive.

Pico

One morning during the end of summer, Mario heard his grandmother talking in a concerned tone to his grandfather, "That ground hog is eating most of our crop, and if you don't do something quick, there won't be any food for the family." Grandfather and grandmother lived with the family and they planted and cared for nearly one acre of various vegetables. Grandfather was a hunter in Italy and was capable of putting fresh game on the table. However, the wood chuck had eluded him to the point of frustration. He was saddened with the situation but somewhat angered from the pressure being applied by grandmother. "What kind of hunter were you in Campodimele, Italia, when you can't even shoot a woodchuck?"

Mario felt compassion for the old man and decided to take matters into his own hands. The next morning, he arose in the dark, and dressed warmly and quietly removed the double-barrel shotgun from the closet. He always accompanied his grandfather and grand uncle when they went hunting. Even though he was very young, they taught the boy the safety and operation of the shotgun.

The large garden was located approximately a quarter mile down a dirt road. The weather was in constant drizzle. Mario thought: The rain is on my side, I can wait and the animal will never know that I'm around. He entered the garden and made his way down a dark row between the pole beans. He took a sturdy bushel, turned it upside down and made himself comfortable. The tempo of the rain accelerated. The boy loved the rain and enjoyed the pitter-patter on the leaves. He loaded the shotgun and placed it in readiness on his lap. The rainy conditions held back the sunlight, the sky lightened very little as Mario scanned the ground from left to right and then from right to left. He felt confident that he was about to solve the family's ground hog problem.

As the sky became lighter he thought to himself: The animal better come around soon. Daylight is breaking fast and my chances will diminish. Besides, I've got to get back before my father wakes up.

His father, Rocco, didn't approve of Mario's interest in hunting. He always advised, mathematics, "You must devote more time to mathematics. It teaches logic."

Mario thought: What the hell does a fourteen-year old know about logic? He kept thinking of his father wanting the son to attend college.

Hero In The Making

As much as the boy was thinking about his father's dream, he kept close attention on the task on hand. He saw movement that appeared to be a blur in between the cabbages. He waited a moment and the blur again appeared from the corner of his eye about five feet from the original sighting.

Mario arose from his sitting position and set the shotgun to his shoulder. He spotted the ground hog in a clear aisle. Mario fired. The blast was deafening and he knew the shot would awaken the neighborhood.

No one had ever fired a shotgun in the middle of town. Mario ran to the spot that he fired at. There was nothing there. He kept the safety off expecting the wounded animal to bolt.

The boy knew that he hit it. Lo and behold, about ten feet ahead and partially hidden lay dead, the biggest ground hog he had ever seen. When it came to guns, Mario was very methodical. He placed the safety on and nudged the animal.

When Mario was satisfied that the ground hog was dead, he picked it up by the tail and hauled it out on to the dirt road. Most of the neighbors were now in the road. They knew about the ground hog and it's eating capabilities.

Everyone congratulated the boy. He loved the feeling. In years to come, people continue to reminisce about Mario and his confrontation with the elusive ground hog. Some old-timers related, "Mario didn't miss that day and he hasn't missed a shot since."

During summer, he would shoot rabbits with a .22 caliber pistol. When someone asked him why, he replied, "The hunting dogs needed training and had to see what the hell they were chasing."

They called him Mr. Hogan. His real name was Salvatore Di Lucci. He was Mario's grand-uncle on his mother's side. The man was a product of the times. He came from a small hamlet of Campodimele located in a high mountain area bordering the Liri Valley. The higher the locations of these communities, the poorer the inhabitants. In fact, most of the families lived in one-room thatched huts. Salvatore visited the town of Pico often and began courting a beautiful young girl that he subsequently married. He left for the United States to make his fortune. The year was 1915 when Salvatore kissed the lovely Alexandria and departed. Thirty-seven years later, the loyal Signora DiLucci is still waiting patiently for his return. He got tagged with the name "Hogan" because he

Pico

was a freewheeling pool player. It was an Italian custom to attach nicknames to colorful characters. He hustled enough suckers to enable him to purchase a pool hall on the second floor of an old building in Little Italy. Some say that Hogan won the pool room in a billiards match. Needless to say, the customers were not your every day pillars of society. The primary method of entertainment was gambling. The customers who weren't shooting pool gambled at cards. They sat around, talked about everything from the old days in Italy to current day gossips and rumors as they smoked and enjoyed their Parodi cigars.

Hogan earned a decent living, but he loved to play the horses. He also loved the many ladies he was acquainted with and spent money on them. Poor Alexandria in Italy, never saw a dime from Salvatore. In reality, he totally forgot that he had a wife. He made money but couldn't hold onto it.

Hogan had a small flat in Rocco's house but there were many occasions when his extracurricular activities prevented him from coming home. However, Hogan loved to hunt small game and kept three hunting dogs at home. He also had his protege, Mario, who was not only his grandnephew but also his best friend. He took Mario under his wing and taught him the ropes. Hogan was molding the boy in his own image. He educated him with the finer arts of pool sharking. Mario also traced his shooting ability to Hogan.

In fact, the student surpassed the mentor. One day, they were both duck hunting. They had double-barrel shotguns as three ducks were approaching. Mario said, "Hogan, I bet you a buck that I can down all three with two shots."

"You've got a bet," countered Hogan.

Mario calmly but alertly waited for two ducks to cross on a line with each other and fired. They both dropped, the third duck was a snap. He shot it as it flew by.

Hogan hollered, "Perbacco," and gave Mario ten dollars instead of one.

Rocco also liked Hogan. Hell, everyone liked him but he was concerned with the deep association between the two. He was also teaching the boy all about women. Hogan brought Mario to a whore house when he was sixteen years old. He informed him, "If you don't get a jump on the girls, you will always be at a disadvantage." Mr. Hogan taught him the five "F's": Find them-Fool them-Feel

Hero In The Making

them-Fuck them-Forget them. He promised the boy, "Follow those rules and it's almost impossible to get into trouble."

Mario laughed to himself: Hogan has so many illegitimate kids that he could probably form a baseball team.

Hogan was happy-go-lucky. He never worried about it.

For some unexplained reason, Mario was never brought to account for his actions. He was supposed to be in school but instead spent most of his time in Uncle Hogan's pool room. The school authorities never checked on the establishment. On other occasions he would be with his uncle at the race track.

Rocco heard that his son was at the track once and asked for an explanation. Mario replied, "Poor Uncle Hogan, he can't see very well and need me to read the racing form for him."

As upset as Rocco was, he burst into laughter, "Are you kidding me? That Cafone can shoot a rabbit in the head while it's running at full speed with three dogs in pursuit."

"Pop, don't forget, Hogan came from the hillbilly hamlet called Campodimele, and can just about barely read and write a little Italian. How can he possibly understand a complicated newspaper like the Morning Telegraph?"

"Hogan is a cry baby, he won't walk in the city because he claims that the cement hurt his feet."

Mario always defended his uncle, "I believe him, Pop, because he will ask someone to drive him wherever he needs to go."

"How do you account for his ability to walk all day when hunting in the woods? He goes from sunup to sundown."

"I don't know, Pop, maybe the ground is softer."

Rocco replied, "How about when it's frozen and hard?" Rocco took a gulp of wine, "Mario, I want you to spend less time with your uncle. Look for a job, try to do something with your life."

Mario shrugged, "What is the use, I'll be going into the service soon."

There was a knock on the door and Emilio Gazzapella entered. He was a friend of both Rocco and Hogan.

Rocco went down into the cool cellar to fill a jar with wine.

Emilio asked Carmela, "I thought that Mario had the job of getting the wine."

"Not anymore," she answered. "A few weeks ago, Mario left the spigot ever so slightly open and about ten gallons of the wine spilled on the floor."

Pico

Emilio looked at Mario, "That is a cardinal sin."

Carmela looked at her son and motioned a chopping gesture. "Your father nearly had a stroke. You wasted the thing that he cherishes the most in life. If you were home, he would have killed you."

Mario replied, "I'll tell you one thing, Pop doesn't bug me anymore about the wine. He goes for it himself."

Camela put her arms around her son, "You devious boy, I know that you did it on purpose."

"Yeah Mom. How about the time Pop came home drunk, fell down outside and slept in the snow?"

"No one knew he was there because it was dark."

Mario continued as Emilio shrugged, "You were so angry that you went down the cellar and broke open the barrel with a hatchet."

Emilio said, "How is that possible, the barrel is made of oak?"

"She hit the section that had the spigot."

Carmela retorted, "It wasn't only because he was drunk. The reason was that he always blamed me for his misfortune. He continuously carried on a verbal attack on my mother and also my father. When he got really carried away, he would include the Madonna."

Mario laughed, "My mother here was very brave, she destroyed the barrel when Pop was in bed with pneumonia. The doctor said he was lucky that he was loaded with alcohol or he would have died."

Carmela replied, irritated, "The doctor, there is another one. He should know, he is always drunk. Besides, if your Pop wasn't drunk, he wouldn't be lying in the snow and would never have gotten sick."

"Brilliant deduction, Mom, you know geometry."

She playfully slapped him on the side of the head. Carmela explained, "Rocco has mellowed. When he was younger, he always was promising someone a punch on the forehead."

Mario laughed, "He always kept his promise."

Rocco came from the cellar with the jar of wine. "Boy it's cold down there."

Mario winked at Emilio, "Pop, do you remember when Mom went on a rampage and busted up your wine barrel?"

He waved a hand in mockery, "She didn't accomplish anything, the barrel was empty and old enough to discard, anyway."

Hero In The Making

Carmela walked away without responding.

Mario looked out the window and observed that Uncle Hogan was approaching through the snow. "Mom, here comes your uncle."

She replied sarcastically, "Give him a big salute for me."

Rocco laughed, "Why are you always picking on the poor guy?"

Hogan entered without knocking, "If this keeps up, I'm moving to Florida and make new friends with the smart Jews."

Rocco poured him a glass of wine, "How are the horses?"

Hogan replied, "They even ran in this cold weather. Ask the bookie Emilio, I lost a parlay by a nose."

"The jockey gave the horse a bad ride."

Emilio was ready to counter, but let it pass.

Carmela, as on numerous occasions, began to demand of her uncle, "When are you going to quit playing the horses?"

The old man was over sixty years old. "It's my business, I know what I'm doing."

They all liked Hogan, but took great pleasure in giving him a hard time.

"He'll quit when a snowball fails to melt in hell," Emilio said.

Mario, as usual, sided with Hogan, "Mom, lay off, he doesn't bother anyone, the horses keep him occupied, it's therapy."

She threw her hands up in the air, "Between the horses, pool sharking and all his other vices, why doesn't he think about his poor wife in Italy?"

Emilio changed the conversation, "Let's go hunting in the morning."

Hogan replied in the affirmative as he held a glass of wine at arms length and declared, "I'm Mr. Hogan and everyone respects me."

Emilio Gazzapella was an ex-jockey and a full time bootlegger of home-made wine. When some ran out of it, Emilio was there to sell some half-decent wine at one dollar per gallon. He was also a part-time bookie who didn't know his ass from his elbow. Most of all, he was a very close friend to Hogan but one would never know it because they argued on every topic imaginable. If Hogan pointed in one direction and signified north, Emilio would point in the opposite direction and declare that location as being north. The problem would always occur when they were hunting in an unfamiliar area. "Where is the car?" Emilio would point one way

Pico

and Hogan, of course, would point to another. They each headed in opposite directions and miraculously both converge on the car that was located to the west!

Emilio was afraid of violating the smallest of hunting laws. Hogan didn't care about any laws. In many ways, Emilio was naive. They were hunting one day when they had no luck at all. Hogan was both disappointed and angry. They were in a very thick area of the woods, heavy with brush and briars when they came upon a stone wall that indicated a property line.

The section was posted "No Hunting and Trespassing." Hogan could care less, he directed the dogs into the forbidden property. Emilio had lost his glasses and could barely see. In fact, he wanted to go home and call it a day. Hogan said he wasn't leaving until he got a rabbit.

Emilio wouldn't cross over the wall. Hogan came onto the scene, "What the hell is the matter with you? The dogs are chasing a rabbit and you're standing here like a jackass."

Emilio pointed at the sign and said, "Can't you see? It's all posted."

Hogan looked at the sign, thought a few seconds, and replied, "It reads NO PARKING. Do you have a car here?"

"No, I don't have a car here!"

They both crossed into the posted land. Without realizing it, it added up to a comedy act.

Mario reasoned that it would be fun to hunt with the Hogan/Emilio combination. Mario said, "If you guys hunt nearby behind the high school, I'll meet you there."

"There are a lot of rabbits there because of the farms, but the briars are too thick," Hogan said.

Mario announced, "Don't worry about it, I'm taking my horse Sparky. I'll show you how to flush the game."

Emilio looked suspiciously, "Are you crazy?"

Hogan sided with Mario, "I think it's a great idea."

"I think it is against the law," Emilio said.

Rocco took another sip of wine, "I don't think it's against the law. In England, they hunt foxes on horseback and also use dogs to chase them."

Hogan took a sip of wine for reinforcement. "Emilio, I can't figure you out. You book horses which is a crime. You bootleg

Hero In The Making

wine which can land you in jail but you are afraid of some stupid hunting laws which can only result in a fine."

Emilio continued with a drink of wine and shrugged.

Hogan continued, "I think that you must be a Calabreze."

"I'm not a Calabreze."

Hogan laughed, "I rest my case."

It was approximately 5:45 in the morning when Emilio approached with his car. Hogan's dogs began barking.

They recognized the sound of his car from a block away and reacted by barking in anticipation of going hunting. Rocco and Hogan were drinking coffee royal as Emilio came quietly to the door. He joined him in the combination of espresso coffee and anisette. Hogan filled a thermos bottle and added the anisette. Rocco didn't care for hunting that much, but he loved wild mushrooms and wanted to check the area.

Mario was sleeping. Hogan said, "I better wake up my Nipota so he can meet us."

Rocco waved him off, "Leave him alone, maybe he'll forget to come. I think it's too cold for him to ride in this weather."

Gazzapella had a black Cadillac. All bookies had black Cadillacs. It was a status symbol. In addition, it was an important car for a funeral procession. Many years ago, Emilio did well as a horse jockey. However, they couldn't leave well enough alone.

Hogan was always trying to beat the system. Emilio was riding a longshot that broke into the lead and ultimately would tire and finish last. Hogan emulated the evaluation of a trainer, "There is nothing wrong with that horse because after the race he looks like he didn't even work up a sweat. All the lazy bastard needs is an electrical jolt from a small battery."

Emilio didn't want any part of the scam, but Hogan kept brain washing him and finally convinced Gazzapella that he, Hogan, would put up the money and buy the battery. The races were being run at a race track called Pascoag.

Hogan placed $200 on the horse to win which, in those days, was a ton of money. The horse took the lead in the six-furlong race. Entering the far turn, he usually began to tire but today, the horse stayed in the lead. Hogan noticed the tail fluttering in the breeze. "Good boy, Emilio, you kept his mind on business."

Coming into the stretch, Emilio must have given him a good jolt because the horse opened up a five-length lead and coasted

Pico

home. Hogan felt like he died happily and went to heaven. At forty-to-one they had $8000-win coming. As Emilio was weighing out on the jockey scales, a judge noticed a bulge in his sleeve. Emilio was so careless that he didn't discard the battery after the race. The horse was disqualified.

Emilio was barred for life. Hogan didn't talk to him for over a month. He then reasoned that perhaps GOD short-changed Emilio with brains. Hogan couldn't blame him but reminded him nearly every day of his life. In the event that Hogan didn't bring up the subject there was always someone else around to instigate an argument. Emilio blamed Hogan for his abbreviated jockey career so they were even. Hogan lost $4000, Emilio lost $4000 and his profession.

They arrived at the hunting grounds as daylight was breaking. It was a gray overcast day, cold and windy. The distance was approximately one mile from Rocco's house. They unloaded the dogs from the trunk and allowed them to run into the brush. Rocco took great pleasure in instigating an argument between Hogan and Emilio.

Hogan ordered, "Emilio, get the dogs into the briars and go in there with them so they will stay."

Emilio gave him a dirty look as if to express: Why don't you go into the briars yourself?

Hogan continued, "And don't screw up like you did that day at Pascoag Race Track."

Emilio gave him a dirty look, "I feel like leaving you here and going home. Come on, Rocco."

Hogan walked away to allow Emilio to cool off.

Rocco knew that he wasn't going anywhere and soon spurred them on, "But who was the mastermind of the operation?"

"That is a ridiculous question. Between us, I'm the only one with a mind," Hogan said proudly.

Emilio countered, "Why don't you ask him where he got the $200? You know where he got the money. He cheated in a pool game." Hogan didn't answer. "I was there, he was playing some big gun from New Jersey," Emilio continued. "The guy was very good, but he had to take a piss. When the guy was out of sight in the toilet, Hogan had a cohort pick up a corner of the table and he placed a book of matches under one leg."

Rocco said, "Then the table wasn't level."

Hero In The Making

"Right, they were so evenly matched that the king here could have lost and you know Hogan, he doesn't lose at pool," Emilio said.

Hogan beamed, "True story but I never considered it cheating.

"Well, what in the hell do you call it?"

Hogan replied rapidly, "Home field advantage." Hogan was warming up, "Can you possibly imagine this clown? I give him a battery that makes his mount the new Man-O-War. My brain power should be investigated, the trainer didn't have the ability to make the horse a winner and I did."

Rocco changed the pace, "We've been here for over an hour and not a bark out of three dogs."

"What do you expect? Emilio won't go into the briars and neither will the dogs," Hogan said.

Emilio gave Hogan a dirty look as Rocco laughed, "They're your dogs. You go into the briars. They will follow their dumb master."

Hogan went partially into the thicket and coaxed the dogs in. He turned, "Why didn't you throw away the fucking battery?"

Emilio walked away without answering. He was looking for a good position in the event the dogs flushed out some game.

Rocco thought with amusement: It's a blind leading the blind. They are both mixed up. The story takes on a new flavor every time it comes up.

Finally, the dogs started barking as a rabbit broke into a small clearing in front of Hogan. He fired and missed. "Rocco, run to the stone wall ahead, the rabbits from here usually make a wide turn and lose themselves in the loose rocks." Hogan was taking charge. "Emilio, go to the left in the bushes and wait there."

"Go to hell, why should I get stuck where I can't see anything?" Emilio cried.

Hogan retorted, "So what? You can't hear anything either. Besides, I'm the boss."

Rocco was having a ball. There was no way they could pay attention to hunting.

Emilio signaled to Rocco, "The big boss missed a shot."

"What happened Hogan?" asked Rocco.

"My glasses fogged up, I couldn't see too well," explained Hogan.

Pico

"What a bunch of bullshit, this cafone has an excuse for everything."

They were making very little progress. The weather was not suitable for hunting. The ground was covered with snow and in the event that a rabbit moved around, the wind would blow away the scent. Hogan said that snow kept the rabbit's feet cold and consequently, the scent was limited. The dogs appeared to be useless.

Mario arrived with his horse who displayed a semblance of friskiness. He got off his spirited animal and Hogan poured him a cup of coffee. "Ah, this hits the spot. Well, how many rabbits did you guys bag?"

Emilio said, "Big time Hogan had an easy shot and missed."

"No kidding, that's unusual, I bet your glasses fogged up," Mario said.

Rocco had left to look for mushrooms. Next to wine, wild mushrooms were his favorite. In fact, he loved wild rabbit with mushrooms and added wine to complement the cooking. Rocco had a knack for this.

Mario took control and directed each of them to take opposite positions in the field so they had a view of the island configuration of the large briar patch. Mario climbed on his charger, "Get ready, I'm going to trample over the briars. I'll flush the game out." He began singing Santa Lucia as he created havoc in the thicket area. Within a period of a half-hour, they shot and bagged seven rabbits and a pheasant. Mario got off his horse and asked Hogan for another cup of coffee royal, "It is still cold as hell".

Rocco came back with some oyster mushrooms that he picked from a tree. He had improvised his jacket as a sack. Hogan gave him a cup of the still hot coffee.

Emilio said, "Where is mine?"

Hogan replied, "Sorry, there's no more."

There was laughter as Mario climbed aboard Sparky and left. "You guys have enough game. Now, I better get out of here before I get into trouble."

Hogan asked, "Hey, Nipota, do you know what George Washington said when he crossed the Delaware River?"

"No, what did he say?"

"Fa un Gatsa da frida."

Mario laughed, "Hogan, you may be my granduncle but I've got to tell you, you're full of bullshit."

Emilio harped, "The bastard may be telling the truth because he's old enough to have been there."

"Hey, Hogan, give those worthless dogs some exercise." Mario rode off as his white scarf fluttered in the wind.

3

An Italian Wake

The funeral home was a large two-story remodeled brick building. Snow adhered to the clinging ivy vines which partly covered a few amber-colored lights. The scene resembled a picture on a Christmas card. The front door opened into a large and long hallway. Three main rooms were utilized as reception areas for visitors to pay their respects for the deceased. The rooms resembled large parlors, as windows were decorated with full-length drapes and the floors were covered with wall-to-wall carpets.

The order of the day was subdued conversation. Seating was on folding chairs which were full. Inside the large doorway, Hogan, Rocco, and Mario were engrossed in quiet conversation. A strong, lingering scent was emitted from freshly cut flowers. It was evident that the deceased had many friends in addition to a large family. Flowers covered the casket and a large portion of the west wall. Mario whispered that at least three cars would be required for the flowers alone. Mario's cousin Gina sat behind a small table. She was busy recording names and accepting donations from the visiting mourners. The procedure was known as the Posta. It was a tradition handed down by generations of Italians. Years ago, most poor families couldn't afford funeral expenses. Thus, the concept of the Posta. Mourners donated money and it was usually enough to cover the financial obligations. In the event that the deceased was deemed wealthy by their standards, they waived the donation.

Another tradition concerned the closeness of the families. It was customary and an honor to cater a meal to the family of the

An Italian Wake

deceased. The entire wake encompassed three or four days, and on this occasion Gina was taking commitments on a first come, first selected basis. The wake was also an opportunity for seldom-seen friends to reacquaint with each other. Needless to say, it was considered a disgrace if a friend did not attend the proceedings. In fact, close friends came every day.

The casket was located at the far end of the room and was highlighted by flowering sprays, wreaths and crosses. It was customary that visiting mourners, upon entering, approach the casket and utter a silent prayer for the deceased. They would then offer their condolences to the family who in this situation were located in seats lining the east wall.

The nearest seat to the casket was occupied by the closest family members and then went down along the line in decreasing importance. In this situation, the first two seats were taken by sisters of the deceased. The third by a brother, and so forth. Male survivors were usually receiving condolences while mingling with the crowd. It was apparent that wakes were a specialty of the female members of the family. A unique event occurred and repeated itself that held the curious attention of the seated mourners. Whenever a new person came in to visit the deceased and offered a silent prayer, the crowd became very silent. The person would then rise and approach the nearest kin. People would focus their attention to the upcoming scene. It was indeed, a display of sorrow. Closeness of friendship was directly proportional to the magnitude of the ritual.

A female family member would embrace the solemn visitor and both would become hysterical with grief. The other sister would join in and the hankies came out to wipe the tears. Upon the visitor repeating the procedure down the line, he would return to the sisters and sit. There was always a seat available. The tears soon stopped and one could hear traces of laughter. This was a signal to the seated crowd to partake in conversation, which at times, was noisy. The complete procedure was repeated every time a new mourner entered the room. It happened for the two hours set aside for the mourning period.

Hogan eyed Mario and motioned him to follow. They walked down a set of stairs which led to a large rumpus room. Emilio Gazzapella was standing near a small table surrounded by a small gathering. Mr. Cosentino, the funeral director, had supplied two bottles of Canadian Club whiskey. This was a tradition for the male contingency of the deceased who managed to slip away for a shot of whisky.

Emilio, the bookie, was very important to the scene. He never missed the opportunity to attend a wake. He was clad in a black suit, white shirt and black tie, and was obliged to be sad and looked more like the mortician than a bookmaker.

Being the man in demand, he was the center of attraction. Emilio adapted himself to the occasion. His importance would lead a stranger to believe that he was an emissary of the family. The bookie was totally cognizant of the age of the deceased, the birthday, the exact time of the death, the day that she arrived as an immigrant to America and any other pertinent information that could be translated into a series of three numbers. It was all business and Emilio was required to supply all the associated information.

Hogan tried to get his attention but, Emilio was busy and ignored him. He wanted to bet a series of numbers but Emilio continued to disregard him. He knew that he could pick up Hogan's action later.

"Hey, Cafone," shouted Hogan.

Emilio placed his right wrist in cradle-like fashion into his left arm and raised the combination upwards. "Va Fongu."

Hogan turned to the people around him and announced, "The son of a bitch got no class."

Everybody was soon betting the numbers. The women were not allowed downstairs, but sent the bets down with male friends. The most popular wager was the exact time of the death, followed by the day, month and year. She expired at two-thirty-seven in the afternoon on January twenty-first.

Most of the action was on two-three-seven and one-two-one.

Hogan arrived at his selection of numbers by way of an old reference book, which among other items, offered designated numbers for various types of dreams. In addition, he had the horse racing entries for the next day. His intentions were to associate the name of horses with some characteristics of the dead. He studied the form with concentration and ignored his noisy surroundings. In fact, Hogan wouldn't answer any questions. He was completely absorbed in interpretation and calculations. A few of his pool room friends would try to distract him, but to no avail. The occasion

An Italian Wake

was an opportunity of a lifetime because he was convinced that the deceased had special powers. People would call her and request that she remove the "evil eyes" from a friend who may have a headache. Some were convinced that among her special powers, Hogan was one.

Rocco finally broke his concentration, "Hogan, what great event is taking place for your state of animation?"

He replied with confidence and determination, "I've got to load up on my bets. The old lady cured me of the evil eyes on many occasions, she always took away my headaches."

Emilio overheard him, "You never had headaches, they were hangovers."

"The nerve of this bastard, I'm going to give him a ton of action and he is insulting me," Hogan swore. He turned and paced a few steps, "I've got a mind to take my business to Choo-Choo."

Emilio stuck his jaw outward, "Face it, I'm the only one that will put up with your bullshit. Choo-Choo won't give you the time of day."

Hogan deliberated a few minutes then gave Emilio $50 in numbers action. He spaced the betting to cover every day of the wake. As soon as she was buried, all bets were off.

Emilio put his hand out for the money. Hogan told him that he could take the money out of his winnings. Emilio frowned, "No money, no action."

Hogan's confidence in winning was instrumental in coming up with the money. He was a strong advocate of the numbers. Every Christmas season, he would play one-two-five starting on the fifteenth of December to the twenty-fourth. The number came out, on schedule, three times in the past ten years. Emilio was smoking one of his favorite Parodi cigars and pouring drinks. Mario assumed he must be giving Mr. Cosentino a percentage of the action.

Hogan observed, "He smokes those ropes like they were Camel cigarettes." Hogan came up with the money and placed all his bets. He now relaxed and conversed with his friends. Hogan always introduced his friends as his nephews. It appeared that everybody there knew him and kidded him accordingly.

The battery incident came up as well as the story when the Padrino backed him many years ago in a high stakes pool match. Word has it that they made a swing throughout the eastern seaboard.

Pico

Hogan made enough money from his cut to purchase himself a toupee, a set of spats and an automobile. The score happened during the lean depression years and he was considered somewhat wealthy. The pool hustler hired a chauffeur because he never learned how to drive. Easy come, easy go. Within six months, he lost everything to the horses.

"Mario," Rocco said, "I just enlisted your services as a pall bearer."

"Why me? It appears that I'm always being volunteered without being consulted."

Rocco responded with fatherly advice, "People will appreciate the respect you have shown over the years, especially during their time of need. You will always be well liked."

"He's right, everyone looks favorable at a person who gives time to be a pall bearer. Some day, it will all come back," Hogan added.

Mario kidded, "Yeah, if I'm running for politics. Pop, this is about the tenth time I've been a pall bearer. At this pace, it won't be long before I'll have done it a hundred times."

Hogan said proudly, "Then you will be able to run for office"

Rocco placed his arm around his son, "Do it for me."

"Okay, Pop, but next time offer my services as a driver."

The rumpus room became crowded. Hogan was being herded into a corner. "Where the hell did all these people come from?"

Rocco joked and was probably correct, "The priest must have arrived and will be leading the Rosary, so all the men bug out. Wakes are for women. They enjoy the gossip and the praying."

Mario said, "I think I'll go and take part in the Rosary. After all, the deceased is my mother's cousin."

He left as his father commented, "Mario is a little on the wild side, but a good kid."

Rocco spotted a character making his way towards the table smiling. "Hey, Benny," he waved. "Over here." Benny was a medium-sized handsome man that usually flashed a smile. During these solemn occasions, Benny wore his tailor-made suits. He also wore the same at weddings.

Clothes always looked good on him. Whenever Benny attended a wedding or a wake, the old ladies would agree that Benny was the best dressed person in town. The fact was, he had only one suit!

An Italian Wake

"Rocco, you son of a bitch, where the hell have you been keeping yourself?" asked Benny. It was only idle conversation because they usually saw each other every day. He approached as Rocco playfully placed an arm around his neck pretending to choke him. "Watch out for the suit!" Benny cried.

Rocco announced what nearly everyone knew, "This bastard has banged nearly every broad on the loose. In fact, if you hold a snake in the proper position, he will take a chance and try to hump it."

Benny raised both hands in denial, "Don't exaggerate, you've got me confused with Gaetano. He is the real hound. Why, that bastard even says hello to little girls. When I told Gaetano that that was improper, he would reply casually, 'They will grow up some day.'"

Benny anticipated what was coming so he tried to change the conversation to the missing Gaetano, "How about when he would watch the daughters of Charlie Chickpea undress?"

Hogan helped him, "You mean the time that Gaetano was looking in their second story window from a tree branch?"

Benny figured that he was out of the woods, "The branch broke and Gaetano screamed as he hit the ground. The firemen arrived to revive him and the crazy bastard jumped up with an open pocket-knife in his hand. He wanted to stab one of them."

Hogan said, "He had good reason because it was after midnight and the sirens drew a big crowd of neighbors."

"Oh," said Benny surprised, "I always wondered why he wanted to knife the fireman. He is fortunate that his wife didn't cut his agates because the branch he broke yielded the most fruit from her pear tree."

Everybody broke out laughing. Rocco wasn't about to let Benny home free. "Bullshit, Benny, how about the time you picked up that beautiful girl somewhere in Cranston. You drove her in your truck to the abandoned gravel pits."

"Come on, Rocco," Benny begged. "Every time I come to a wake, somebody always manages to bring up an embarrassing episode. Hell, what happened to me could have happened to anyone in this room."

Rocco was holding court and had everyone's attention, "Benny here is a strong, normal red-blooded Italo-American boy. He couldn't wait to get started and began kissing the broad."

Pico

Benny tried to stop the story, "Come on, Rocco, cut the shit."

Rocco paid no attention, he continued, "When the great one here, gets down to business and starts to undress her, he placed a hand down between her legs and got the surprise of his life. He added some emphasis, "Hello, what do we have here? Can you believe, a queer dressed in flashy woman's clothes? He had long natural hair and was wearing make-up. The expert here was completely taken!"

As usual, everyone laughed and some made derogatory remarks in Italian that sounded musical. Benny yelled over the voices of the loud crowd, "I grabbed the son of a bitch and dragged him to the dirty pond by the gravel pit."

Rocco inquired, "And then what?"

"I threw him in. It was about a twenty-foot drop. He kept yelling, 'Please don't hurt me,' but he was more concerned with his dress. Can you imagine, I'm belting the shit out of him and he's worried about his dress!"

Hogan asked, "What happened to him after that?"

"I didn't give a fuck, but he had a fifteen mile walk back to where I picked him up." Benny kidded, "I got to tell all you guys one thing, that queer was a beautiful looking girl."

"I bet you all that, over the years, he fooled a lot of people."

Benny added, "He attended Sunday Mass dressed that way."

Rocco teased, "How do you know so much about the queer's current events?"

"Fuck you, Rocco, what the hell, are you writing a book or what?"

Someone in the gathering asked, "Benny, you are the main topic of the conversation. How about the time that you lifted that bulldozer from the City and stashed it away in the woods?"

Benny was much more comfortable with this line of discussion, "I would rent it out to shady friends at half-price."

Rocco joined in, "This asino once went around stealing all the picnic tables from the rest areas and sold them cheap to his friends."

Benny said laughing, "Times were tough, the tables are scattered in back yards all over town. Someone called out, 'Benny, I need a table, how do I put in the order?' And I say I can't help you, the state people now chain the tables to trees."

Benny was an opportunist. One year, near the end of the great depression, a hurricane hit the area. People stayed in their homes,

An Italian Wake

but not Benny. He took his truck and drove around, taking whatever valuables were laying around.

Hogan shook his head in disbelief. Johnny Catalano entered the room asking, "What the hell, are we having a party here? Is this a wake or a wedding reception?" Someone replied, "Fuck you, Cats." Johnny was one rough bastard who didn't take any crap except from his friends who were allowed to banter with him. He appreciated it when his admirers recounted his exploits. Hogan usually gave Johnny a hard time because Cats was a habitue of the pool room. Hogan related the time when the police wanted to take Johnny down to the station for questioning. Cats was playing pool and told them, "Perhaps later." It took seven cops to finally get him into a patrol wagon. Cats could really fight and he put three of them in the hospital. He hardly ever worked, but he always had money. Johnny wasn't a bookie, but he drove around in a well-kept black Cadillac. Hogan said as Johnny was listening, "Years ago, this asino and I were coming from an all-night poker game. A police car starts to follow us and instead of slowing down, Johnny picks up speed. The cop puts on his flashing lights and gives us the siren. Of course, Cats knew the cop so we pull over. He starts to read Johnny the riot act. Unknown to me, he was pissed at the lawman because he caught him cheating at a Sunday morning crap game."

Benny asked, "What did Cats do?"

"What did he do? He got out of the car and beat the living shit out of him. Adding insult to injury, he took his service revolver and threw it off a bridge and into a river," Hogan assured.

Rocco added, "I believe it about Johnny. He once busted a chair over a guy's back at a poker game because the guy looked at him funny."

Johnny Catalano took a deep drag on his cigar and blew out a perfect oval, "Rocco, look who's talking. You're the top scrapper in the whole area."

"I guess, when I was young, I could handle myself pretty good," Rocco conceded.

Johnny was referring to the time that Rocco had a little misunderstanding at a local wedding. Three young men from out-of-town were bothering a shy girl at a dance. She ignored their advances but they kept bothering her until Rocco came to the rescue. He warned them to stop pestering her and left to inquire about the three men. He discovered that one was a professional fighter, another one was a gangster and the third was nothing to worry about. The outsiders continued to bother her.

Rocco approached and threw them off guard by offering to buy them a drink. The next thing that happened was he hit the fighter with a tremendous shot that shattered his jaw. With lightning speed, he picked up the gangster and tossed him down the stairs. The third guy ran away.

Hogan said, "I was there, Rocco took off and hid in the priest's rectory. When the cops came, the priest told them he didn't see Mr. Calcagni."

Rocco didn't comment at all except that he now needed to make an appearance upstairs. Benny and Hogan followed, the fun was over and it was time to participate at the conclusion of the Rosary. As they were going up, Mario came back downstairs. Johnny commented, "Mario, I hear that you're going to be a pall bearer."

"Yes I am, why?"

Cats replied wryly, "Nothing, except that sometimes those caskets get heavy." Mario shrugged as he and Johnny joined in idle conversation. "Kid, I need another guy to help me with a little job, would you be interested?"

"What kind of a job?" Mario inquired.

Johnny Cats knew that he could divulge this information to Mario without concern. He trusted Mario, it didn't matter whether he accepted or declined. "There's a restaurant in Newport that does a tremendous amount of business. You know, the type of establishment that has a grocery deli in front and an eatery in the back." Mario was listening as Cats continued, "The owner never deposits money in the bank, he hates bankers because they screwed him during the Depression. The dumb Cafone has a big heavy safe in a small, rear office that is loaded."

Mario didn't want any part of robbery but kept listening because Johnny commanded respect.

Cats continued, "I have a girlfriend who works for the owner. She gave me the whole layout. It's too good to be true."

" I don't know, Johnny. Girls talk too much, they have a drink and start bragging."

"Not this doll, she doesn't drink and I've known her for a long time."

An Italian Wake

Mario wanted to walk away without hurting Johnny's feelings. Instead, he played along. "How does this girl know so much without arousing suspicion?"

"No problem, she gives the owner a blow job a couple of times a week."

Mario laughed, "She does that and she is your girl?"

"Kid, business is business, besides, she is just a another putana."

Mario shrugged, "Why are you asking me?"

"A couple of reasons. One, the safe is very heavy and two, you know how to use a cutting torch. You, I and crazy Joey can move that safe out of there in five minutes."

Mario raised his eyebrows in disbelief. "Joey is a drunk and unreliable to boot."

"You are right but he is a strong Santo Antonio and when he thinks that he is important, he keeps his mouth shut. I'll tell you more, he is so stupid that we won't have to give him much money. Besides, he has an old panel truck that we can use." Mario was still contemplating an opening to end the discussion as Cats continued, "The girl lifted a key to the office over a month ago and the owner doesn't even know about it. I've been in the office myself after work."

Mario's eyes lit up for the first time, "Are you kidding me?"

"Of course not. Hell, I've been observing that safe on at least seven different occasions."

"Why would you take a chance of going in there without getting the safe out?"

"It's perfectly secure because it's the only place I have the privacy to fuck my girl!"

Mario became interested and thought to himself: This isn't even a risk. He was thinking and Johnny knew he had him sold. Mario asked, "What about the cutting equipment?"

Johnny replied, "Joey stole what we need out of a welding shop near Boston."

" You know, Cats, that equipment can be traced."

"Hell, Joey lifted it over a month ago and there is no trace."

Mario shook Johnny's hand, "Okay, count me in."

4
Setback To A Career

Mario was riding his horse, Sparky, up and down the street. He wore his riding boots and had a white scarf wrapped around his neck. The animal was slow afoot, but Mario didn't care. He treated the horse as if it were a charger from Camelot.

He would, at times, ride on the sidewalk. He loved to sing arias and knew many Italian songs. Mario was an excellent tenor and took pride in displaying his talent by bellowing out early Sunday mornings as the neighborhood slept. He would open his window and emulate Enrico Caruso.

The neighbors were infuriated. Between the horse and his singing, they were at the end of their collective rope. Everybody loved Mario but he was becoming a nuisance.

In the spring of 1942, Mario was waiting to be drafted into the service. Whenever he caused a minor problem, his father would comment, "Uncle Sam will straighten him out." There was usually a friend around who would answer, "Mario will straighten Uncle Sam out."

Mario always took shortcuts when riding home. Benny, the Hound, Parenti lived in a house that bordered on two parallel streets and had beautiful, sturdy hedges as a natural fence. Mario was always ready to test the jumping ability of his graceful charger. Sparky never made a successful jump over the hedges and consistently plowed through. After completing the shortcut, Mario would turn to onlookers and proudly announce, "Did you see this son of a bitch jump those hedges?" One concerned observer

Setback To A Career

replied, "One of these days, Benny is going to shoot your ass off that plow horse."

"Fuck Benny, he's a friend of mine."

Mario had a distinctive command of four-letter words. He had mastered the utilization of them. He caused an adverse reaction when using the word. People actually laughed.

Whenever Mario met Benny, he immediately asked, "Did you and that plow horse trample my hedges again?" Mario never replied to the question. He would answer, "Hey, Benny, how's the wife and kids? Have you been out picking mushrooms lately?" Benny would shake his head and walk away. A friend once asked Mario why he gave Benny the runaround. He replied, "Benny is one of the best fucking guys in town. All he needs to do is stop chasing all the available women. That way, he will have more time to keep an eye on his hedges." In reality, Mario was after the same women.

One day, the old-timers were playing bocce on a grassy lawn of the local park. As they were in a dispute over a game, which was a normal occurrence, Mario pranced his charger through and purposely scattered the balls. They cursed him in Italian and he gave them his standard answer, "Fuck you." One burly participant picked up a bocce ball and flung it at Mario. The ball hit the horse squarely on the flank. The plow animal took off like a bat out of hell.

They all laughed. One old-timer said it all, "Mario is one crazy bastard but a lot of fun."

Someone asked, "Mario, when are you going to get a job?"

"Iggy, I'm a singer. As long as people get married, I do okay."

"Mario, why don't you sing in church?"

"The priest doesn't like me."

"He could sure use you in the choir as the lead singer, all you need to do is stop swearing."

Mario pulled the reins on Sparky and backed him up as he replied, "The priest doesn't care about my swearing, that's not the problem. The good padre holds a grudge. He is still angry with me since the days that I was an altar boy." Mario revealed, "When he tapped for more wine during mass, I only poured a few drops. Hell, he was drunk most of the time and I was only looking out for his best interest."

"Mario, you're crazy, you should have more respect for the priest."

Pico

His charger was now prancing in place. "One day, after Mass, he caught me and slapped me on the side of the head. After all I did for him, he had the nerve to slap me. He once got drunk at my father's house with all his paisanos and I had the job of taking him home. I lined the drunk up with the front door, rang the bell and ran like hell. The priest was new to the congregation, I think he recently arrived from Italy. He must have caught holy hell because he didn't join the usual get-together at my father's house."

Mario patted Sparky on the side of the neck. Iggy asked, "Then, am I to understand, that you are never going to sing in the choir?"

"No, not until he apologizes to me."

"Mario, you're are a character. As great as you sing, he won't allow you in his choir. You'll be whistling in the wind if you expect an apology from a priest."

Mario replied casually, "I don't give a fuck, I'll sing in the Portuguese church."

Benny who was sitting on a bench heard most of the conversation. He shook his head in disbelief, "Mario, you are good for only two things, singing and shooting."

"Hey, Benny, the hound," Mario hesitated momentarily to get everyone attention, "fuck you." Benny gestured by throwing his hands skyward as he added, "I've also screwed more girls than you and I'm only nineteen-years old. All kidding aside, you better stop fucking your sister-in-law. One of these days, your wife is going to nail you."

Iggy countered, "If his father-in-law finds out, the hound will need to run off and join the Marines."

"Screw all you jerks, I never fucked her."

Mario galloped away with parting words, "Too bad, Benny, I fucked her and she's not bad at all."

"You lying son of a bitch," said Benny.

Children were playing stickball in the street as Mario rode up on his horse and momentarily broke up the game. The kids responded by tossing rocks after them. It didn't bother either Mario or his horse. One of the boys hollered, "Get a job!"

It had rained all day and continued into the night. Johnny and Joey picked up Mario on the corner. Johnny winked at Mario and asked to Joey, "Do you think this fucking iron horse will make it to Newport and back?"

Setback To A Career

Crazy Joey answered with confidence, "It may sound like it's falling apart but this truck is reliable as all hell."

Mario just listened to the two yo-yos.

Cats continued, "This goddamn heap can't get over 30 miles an hour. How much slower will it move after we load that safe?"

Mario couldn't resist, "Is that an algebra problem?"

"What the fuck is algebra?." Joey laughed.

Johnny replied sarcastically, "Are you writing a fucking book or something?"

Mario got back on the thing bothering him, "Don't get me wrong, but aren't we taking a chance with this candidate for Frankie's junk yard?"

"Its all on the spur of the moment, we had something better lined up but crazy Joey got his signals crossed," Johnny replied.

Joey looked ahead without responding. Mario continued, "I have an uneasy feeling about this whole operation. Maybe we should call the whole thing off until we can line up a better truck."

"Mario, I would agree with you a hundred percent, but the owner is going on vacation and he told my girl that he is going to move the money. Tonight is the only night, don't worry, everything will turn out fine. You are a little shaky. It's your first time and a normal reaction. Just take a look at this crazy fucker. Nothing bothers him."

Mario couldn't hold back laughter because Joey thought he was receiving a compliment. Mario was far from relaxed.

They arrived at the destination at one in the morning. The rain appeared to intensify and the area was completely deserted.

Johnny opened the door with the stolen key. He turned on the light and pointed to the big safe, "What did I tell you, as easy as pie." Mario had to admit, the plan was working so far.

They brought along a heavy duty movers dolly along with a couple of crow bars. Mario and Joey lifted the front of the safe with their long levers. Johnny placed a wooden block under the raised safe in order for them to get better leverage to raise it higher. When they had it high enough, Johnny pushed the dolly under the safe. He wedged a two-by-four against the dolly and the boys released the levers. The safe tipped down and landed squarely on the dolly. Johnny sighed, "Just like downtown." They set a makeshift wooden ramp from the truck to the ground. It was constructed of two-by-fours bolted to a thick sheet of plywood. The ramp protruded a few feet into the truck. All three of them pushed and heaved until the safe was nearly into the doorway. Johnny and Mario wedged the dolly/safe with crow bars as Joey backed-up the truck. The safe and ramp both dropped to the floor of the truck. They easily forced the rest of the ramp foward and closed the door. The entire operation took fifteen minutes.

They left the alley and entered the street. Johnny and Joey relaxed and sang even though neither could carry a note.

"How much money do you guys think there is in the safe?" asked Joey. Mario had no idea, he never got around to asking.

Johnny Cats replied, "At least twenty thousand. This is the easiest score that I have ever pulled."

Mario said, "We're not home free yet, this clunker really scares me."

"You are one pessimistic asino. I bring you in on a score that turns out to be a gift and you're giving me the evil eyes," said Johnny.

It appeared that Johnny was looking for a fight, but Mario let it slide. It wasn't so much about the comment, it was the derogatory way he said it. Mario thought to himself: I should be happy over the whole operation but somehow, I wished I never got involved. He glanced at the speedometer and noticed the arrow at thirty miles per hour as the entered the Mount Hope bridge. If they maintained the snail's pace, they would reach the garage in about a half-hour. The two were still singing off-key when they heard a loud bang and the truck slowed and came to an abrupt halt. They were approximately five hundred feet from the toll booth.

Mario hollered, "I knew it! I knew something was going to happen!"

Joey jumped out and checked the problem, "The left rear tire blew out!"

Cats ordered nervously, "Get the fucking jack out and change the tire."

They were fortunate that there was very little traffic at that time of the morning. Mario got the tire out as Joey was fumbling with the Jack. The man in the toll both was becoming agitated because the truck was obstructing traffic.

Setback To A Career

Mario gave Johnny a dirty look, " I was right, this fucking truck has put us in deep shit." Johnny didn't answer as he was trying to help Joey who was having difficulty with the placement of the Jack. Not only was the rain a problem, but the wind on the bridge resembled gale conditions. They raised the truck enough to enable them to remove the wheel.

Mario thought as he wheeled the spare tire near the flat: We may still have a chance.

Just then, a heavy gust of wind swayed the truck. The jack tethered and fell down to the hub. "That's it," Mario was pissed off, "I've had it with you two fuckers."

Johnny responded, "You jinxed this operation from the start, I ought to knock you on your ass."

Mario walked up to him and pointed a finger, "Take this fucking safe and shove it where the sun doesn't shine."

Johnny Cats was enraged, he took a wild swing at Mario and missed.

They went at it fast and furious when a patrol car approached from the toll side of the bridge. Mario was quicker with fast hands and got the best of Johnny and left him struggling over the hood of the truck. The great Johnny Catalano finally met his match. He was bleeding profusely. The police drove up and blocked Mario's path. He was asked, "Where are you going?" Mario thought: I'm through with those two jokers, they can keep the safe for themselves. He answered the police, "I'm going to make a telephone call to locate a jack, the one we have busted."

"Hold on a bit, I think we can call for some help."

Mario had no alternative but to reluctantly walk back to the truck.

"What the hell are you guys fighting about? " asked one of the policemen. There was no reply and Mario surmised that Johnny was contemplating taking a punch at the policeman. If he wasn't so beat up by Mario, he would have.

"What in the hell are you guys hauling in this junkyard heap?"

"Evergreens," replied Johnny.

"What are you men doing out in this weather at two o'clock in the morning?" The cop-in-charge directed his partner to take a look at the cargo. He noticed Johnny Cats was ready to make a move which compelled the policeman to draw his service revolver. "Look here, a big safe that looks nothing like evergreens."

Pico

The three were arrested and brought to the station. The police lieutenant called out from his office, "Bring in that Dago kid, I want to talk to him." He was a big heavy-set man and had his sleeves rolled up as high as they could go. His arms were huge and he used them to intimidate people. Lieutenant Daniels had the reputation of a law man who took great pride in beating information out of a suspect. Mario was aware of his reputation. In fact, Benny once told him a story about super-cop.

The Lieutenant was shacking up with a woman who was married to an even bigger man. Apparently, the beautiful married lady had a few too many speeding tickets. He coerced her into a trade-off. Her husband heard about the arrangement and paid a visit to the police captain. He threatened him with the possible loss of one sick Lieutenant. The captain was convinced by the look in the the man's eyes that the threat could become a reality. Consequently, the super-cop kept his distance from the woman.

Mario sat in a chair facing the lieutenant who continued with his paper work. Mario sat quietly. He knew about the lawman's reputation but was not afraid of him. In fact, Mario wasn't afraid of anyone.

The lieutenant finally looked at him with a fake smile while opening the center drawer of his desk. He removed a claw hammer and tapped it gently in his massive hand and finally spoke, "Another young Dago who has stepped into the big time."

Mario didn't reply. "I have, here in my hand, the remedy to straighten you out." He proceeded to pound the hammer faster and with greater force into his open hand. Mario eyed him with silent curiosity. "Answer me, you Dago bastard, or you are going to be one sorry son of a bitch." Mario had no idea on what he was supposed to reply to. There was no question that necessitated a response. Lt. Daniels left his chair with a vengeance and waved the hammer at Mario, "Answer me or I'm going to plant this hammer on the side of your head."

Mario replied calmly, "Fuck you."

The officer's eyes popped in disbelief and he started to lift the weapon towards the boy's head. Mario shot his left hand upwards and caught the Lieutenant's wrist. He tightened his grip with all his strength and stopped the aggressive motion. Mario twisted Lt. Daniel's right hand and placed two fingers under his bulging eyes,

Setback To A Career

"Don't ever call me Dago or I'll pluck the ugliness out of your head." The lieutenant grimaced with fear as Mario forced a smile, "Do I make myself clear?"

"I can call for help. You're assaulting an officer."

Mario placed his right hand on the lawman's throat, "I don't think you should ask for assistance." His adversary remained frozen as Mario emitted a snapping force at the wrist, which allowed the hammer to drop harmlessly to the floor. He explained, "I'm going to release my grip, I will not reveal this incident to anyone. You can continue to maintain your false image of Mr. Tough Guy. But remember one thing, if you fuck with me, I'll get even with you somewhere down the line."

Mario stared into his eyes to stress the point. When satisfied with an eye response, he released his hold.

The Lieutenant rubbed his wrist and then his throat.

He summoned a policeman in the outer office and instructed, "Throw this fucker in a cell." Mario tried to contain a half-smile as he left the office but also realized that he could easily be shot in the back. The standard excuse being that he tried to escape.

The makeshift prison cell reeked of staleness. He sat on the dirty cot and began reading the writing on the walls. Mario was surprised to see recognizable names of people whose incarcerations were not publicly known. The place was filthy, but he knew that he would be out by morning. His father would be out retaining the best Jewish lawyer available.

He looked out the window towards the library and recalled the many times he visited. He had nothing but time now and consequently layed down on the cot. Mario stared at the ceiling and made out the name of Joe "Tripe" Trippa who wrote, "Welcome to the exclusive club of Cetriolos." Cucumbers. Mario couldn't help but laugh. Tripe was the biggest local thief around and finally attained a status of legitimacy by operating a successful Bar and Grill. He threw it all away when he caught his bartender stealing and proceeded to chop off three of his fingertips with a meat cleaver. Mario talked out loud, "Tripe was sure a trip." It was early afternoon with nothing to do. He didn't want to lie around and sleep and as an option, he began to do push-ups. Mario continued the exercise for nearly three hours. His arms felt heavy as his chest ached with soreness. He reasoned with satisfaction: It all hurts but hurts good.

Pico

He laid on the cot but got restless and decided to continue the same exercise before retiring for the night.

The ceiling also revealed a well-known name. Tony "Fuoco" White. His real name was Tony Bianchi. Tony was the foremost arsonist of the area. He operated a flower nursery as cover for his prime vocation. Mario recalled the day when Tony purchased an old beat-up house and had it moved to the rear of his property. He hired a drunk to do some cosmetic carpentry work.

The place was uninhabitable but Tony conned an insurance man to insure the building. The neighborhood wise guys were making bets on when Fuoco would put his expertise into action.

On Halloween night, it happened.

Mario couldn't help laughing to himself when Rocco informed him, "Poor Tony, he worked so hard and some figlios di putanas burned down his house on Halloween night." Rocco liked Tony and never believed the arson stories. He said, "Tony had respect because he always bought Rocco a drink at the Sons of Italy Club."

"Yeah, Pop, I know all about it, the best drink in the house for Rocco." The sharp guys on the street were so sure that Halloween was the night that Tony selected, that they formed a money pool and drew for time slots. Rocco shook his head in disbelief.

Mario thought to himself: Here I am, laughing at Fuoco and I'm the asino that's in jail.

Tony had once confided to Mario, "I did my first torch job when I was fifteen years old. This kid took my girl from me because he had a car. I retaliated by burning the vehicle."

Mario had egged him on, "Are you bullshitting me?"

"No way kid, always use high test gasoline. Regular is a waste of time, too slow to react."

In later years, Mario recalled asking Tony about Halloween night.

He boasted, "High test, the only way to go," as he handed Mario a cigar.

"Excellent cigars, where did you get them?"

Tony waved him off, "I seldom buy anything. Let me tell you a fact kid, that fucking house was history in less than ten minutes. The authorities couldn't find me for a week because I took a vacation as soon as I lit the fire."

Mario urged him on, "What happened when you returned?"

Setback To A Career

"The bastards tried to give me a hard time, but I proved that I was out of town. The only thing I had in there was a pot belly stove. It turned into a beautiful mound of cast iron slag." He burst out with a roaring laughter, "High test baby, high test."

Mario wondered how in hell he got his name on the ceiling. He doesn't remember Tony ever getting arrested. They're fortunate he didn't return to burn the police station down. Mario lay with his hands clasped behind his head and murmered, "I sure fucked-up."

Rocco was pacing the floor as Carmela started crying, "I told you, that wild uncle of mine would someday get Mario in trouble. If Mario was kept out of that Pool Room, he would never have met that gangster Catalano." Rocco formed his hands as though he were praying and placed them towards the heavens, "They ruined my son, my Mario."

Carmela kept sobbing, "The old wise people have a saying, tell me who you socialize with and I will tell you what kind of person you are."

Rocco stated with a threatening finger at Carmela, "I don't ever want to see your Uncle Hogan around this house anymore. If I ever find him here, I'll throw him out the window. I don't give a damn about him and his mafia pals."

Benny was present during the discussion, "Rocco, is there anything I can do?" Rocco nodded with a negative response. "You know Rocco, I'll do anything for the kid. The word's around that Mario was walking away and that the police nearly let him go."

Carmela bit over the side of her outstretched hand, "It's that Johnny Catalano, I hope someone shoots him."

Benny said, "For all it's worth, I heard that Mario gave him one hell of a beating on that bridge. From what everybody is saying, it is the first time that the great Johnny Catalano ever lost a fight."

Rocco replied disapointedly, "Fighting, guns and robbery will get you nowhere."

Benny agreed, "You're correct but the thing that surprises me is that nobody is talking about a robbery. The fistfight on the bridge will be talked about by future generations and Mario was the victor."

There was something special about Mario Calcagni. He became a hero by disposing of a hungry, destructive ground hog. He gets

Pico

into trouble and yet will be rembered as the young man who put "Cats" Catalano in his place.

Benny thought silently: Mario was one of those special people born to become famous.

Within a few days, Rocco got Mario out on bail. He had to use the house for collateral. When Mario came home, he was subdued and kept to himself. He never started a conversation with his parents and only answered when asked. Rocco fortified himself by drinking more wine. The situation was now different. Previously, he drank wine with meals and socially. Now, he was drinking wine without food. Mario knew that soon this new situation could turn his father into an alcoholic. He also knew that he was responsible and hoped that time will be a healing factor. Carmela would forgive and ultimately forget. There was something about mothers in regards to their sons. They could do no wrong. Mario pleaded guilty for the felony and was placed on probation. The judge reasoned was that Mario was a first time offender and that he was intending to stay away from trouble. The judge didn't say, but he was elated with the outcome of the fistfight on the bridge. Johnny and Joey were sentenced to serve time in jail.

Uncle Sam would no longer have the responsibility to straighten Mario out because he was now classified as undesirable and exempt from the draft.

Benny observed as Mario was reading the notification for the second time. Mario spoke with disgust, "Screw the Army, screw the Navy and fuck the Marines."

He brooded and kept to himself. His happy-go-lucky days were over, at least in his own mind.

For some unexplained reason, his friends and neighbors thought the same way. He knew and began accepting the fact that his life had changed.

He thought to himself: Now I can be like the Sisto brothers. They're convicted felons and the Service doesn't want them. Come to think about it, they are doing very well operating an auto salvage yard.

Mario answered his own thoughts: Screw the Sisto brothers, they're not worth the crap they wallow in.

The following days were spent in seclusion. Friends called, but Mario refused to take the telephone calls. He no longer

Setback To A Career

socialized. His parents accepted the fact and saw only the good side of the situation. They had this one child and he would not be killed in the war. Their Mario would be safe within the borders of the United States.

5
Second Chance

A few weeks after the robbery, Rocco met an old friend, Patrick Morgan, a retired Navy Commander in World War I and a respected engineer. Patrick was never asked what field of engineering he specialized in because he excelled in all aspects of engineering. The commander had friends in prominent positions, especially in the arena of local Irish politics. Patrick had a high regard for Mario, but he was disappointed about his recent activities. However, he saw some good qualities in the boy. Rocco ordered two shots of brandy in the local Irish bar.

"Patrick, I guess that by now, you've heard about Mario."

"Yes. You know that I think the world of that boy."

"I know you do, Patrick. What can we do? Can anyone do anything about the present situation?"

The Commander sat back and lit a cigarette, "I have already given it some thought. It will depend on Mario."

"What do you have in mind?"

"Rocco, I don't want to get your hopes up too high so I'm not going to tell you anything."

"Are you going to help him?"

"Of course, I am. At least, I'm going to try." He placed a hand over Rocco's hand and smiled assuredly, "Be patient."

Patrick Morgan had a heart condition and wasn't allowed by his wife to drive his car. She was terrified of him driving and consequently he had a transportation problem. One early summer day, Mario was sitting alone at the park watching a sandlot baseball

Second Chance

game. He was laying on the grass feeling sorry for himself. Patrick Morgan approached and sat next to him, "Hello, young fellow."

"Hi, Patrick, how are you doing?"

"Fine, except for the fact that my wife won't allow me to drive a car because of my heart condition."

"Etta is only looking out for your best interest. You know that she's right."

"I suppose so, but if I don't keep busy I'll kick the bucket anyway."

Mario thought the world of Patrick and smiled for the first time in a month. The commander got right to the point, "I need a favor."

"Sure, Patrick, what kind of a favor?"

"I want you to drive me around and help me with my consulting work."

"Are you offering me a job? Me, Mario the gangster?"

"You underestimate yourself, kid. Pull yourself together and be at my house at nine in the morning."

Mario hesitated momentarily and seeing the sincerity in Patrick's eyes replied, "Yes, sir."

Mario rose from his prone position on the grass and helped his future mentor to his feet as a foul ball came in his direction. Mario caught it barehanded and fired it on a line to the unexpected center fielder. Patrick said, "That brings back memories of the day in a high school game when you threw that runner out at home plate."

"Yes, Patrick, I recall it clearly, center field to the catcher on a line. I had a great arm."

"You still have a great arm."

They were walking away from the game as Mario quizzed, "Why are you helping me?"

"You've got it backwards. You are helping me, and who knows, perhaps, I can make an engineer out of you."

The next morning, Mario arrived before nine o'clock. Patrick wasn't ready. In fact, Mario would find out in the future that the Commander moved slow in the morning.

His wife, Etta, always prepared a soft-boiled egg along with toast and coffee. The meager breakfast was a standard fare without any deviation. She was constantly on his case to move along faster, but to no avail.

Pico

The routine was a normal occurrence.

He would reply to Etta's urging, "Okay, Mama, everything is going to be fine." Mario could never understand why he called her Mama because they never had children. They would finally leave for a job site and Patrick would direct his driver to stop at the nearest coffee house for coffee and donuts. Mario made a casual observation over the ensuing few months: Navy people sure drink a lot of coffee.

Etta was happy and comfortable with the new situation because the commander was in good hands. They knew Mario as a child and both had a genuine interest in him. In fact, there were times when he felt that they had found the son they never had.

Patrick, like many Irishmen, was a devout Catholic. On Sunday mornings, he passed the collection basket at Mass. In the event that an altar boy didn't show, he filled in. The commander was active in the Catholic Youth Association. Patrick was the neighborhood scoutmaster and organized baseball pick-up games during summer evenings. However, creeping age and the heart condition slowed him down considerably. He taught Mario surveying, metal fabrication, report writing, heat engineering etc. Patrick was the private mentor and Mario was the attentive apprentice. The mentor always repeated the advice: Read, read and read some more.

They had a common bond. Mario had an ability for getting along very well with older people and Patrick Morgan could easily relate with the younger generation. The respect was always there. Over the years, Mario never used a four letter word in Patrick's presence. He protected the old man as one would protect a grandfather. One day they were having a meeting over coffee in a diner. An owner of a construction company approached the table and began to get belligerent with Patrick over a previous consulting job. He complained that he was overcharged. Mario knew that Patrick never charged enough for his work and became concerned with the proceedings because of the heart problem. Mario broke the escalating discussion, "Sir, I know all about the incident that you are referring to, can we go outside?" Before the angry man could reply, Mario added, " I can make up the difference if you quiet down." He relaxed and agreed. Mario turned to Patrick, "Be calm, I'll be right back."

Second Chance

They walked outside and around a corner. Mario wasted no time. He hit him with his left hand with a tremendous punch squarely on the side of the face. He was preparing to follow up with a right, but it wasn't necessary. His adversary didn't have a chance. He was out cold.

Mario went back in, sat down and continued with his coffee. Patrick asked, "Is everything all right?"

"Of course, I only kept my promise by making up the difference." The commander knew very well what happened.

Rocco was happy and appreciative with Mario's progress. He drank less wine and returned to his normal way of life. Everybody was satisfied with Mario except himself. He couldn't get comfortable with himself because he had that dark cloud following him around: A felony.

Mario also lost interest in chasing the girls.

One morning, as they stopped for their customary coffee, Patrick asked, "What's wrong, Mario?"

He hesitated and fumbled with a spoon by tapping the cup. "I don't know, Patrick. I certainly feel good about accomplishments but I have no goal. I'm not fooling myself, that damn felony fiasco is going to follow me the rest of my life. Can you imagine me telling my children, if I ever have any, that I'm a convicted criminal?"

The commander was waiting this opportunity. Mario was opening the door and Patrick was ready. The time was ripe. "Mario, I've given your problem a great deal of thought. I believe there is a way out." Mario was listening curiously. "You, my son, are at the crossroad of life. You can take the easy way out by hanging around the poolroom with a bunch of undesirables. Sure, you will have a lot of fun and probably make a lot of money. The chances are that you will be a big gun like the Padrino. You certainly have the potiental and are well liked." Mario continued to listen with interest. "Between Hogan's influence and your exploits on the bridge, I'm amazed that you haven't been contacted."

Mario replied, "They are afraid of my father."

"I know all about the Padrino, all the good things he has done."

The Padrino was known to send a doctor to a family that could not afford medical attention. He also distributed food to needy families. He gave gifts to children on Christmas, a genuine Santa Claus.

Pico

Patrick pointed an index finger at Mario to emphasize the advice, "But remember, he is a sophisticated crime figure who has done time in the past and will probably do time in the future."

Mario replied while leaning back in the booth, "I don't want to be a Padrino. Among other disadvantages, they don't live long."

Patrick accepted more coffee from the waitress and waited for her to leave, "I'll come right to the point. You must take the long and difficult road."

Mario asked, somewhat surprised, "You mean there is a way out?" Before Patrick could reply, Mario added, "I'll do anything to wipe out that felony."

The big Irishman said, "I believe it. You have certainly proven your character to me." Patrick was now on a roll. He had formulated a plan for Mario even before he was approached by Rocco. "First of all, I will write a petition. I have aligned some influential friends who have a lot of faith in you. We will canvass the entire area for signatures in your behalf to get you into the military service."

"What's the service got to do with it?"

"That is the only way to erase the felony rap. Of course, you will be required to obtain an honorable discharge."

"You mean to sit there and tell me that a petition will get me into the service?"

"Not entirely, there is one other obstacle to clear."

Mario eyed him with intense interest. He didn't need to ask. The suspense in his gaze did the questioning.

"The Attorney General of the State has the final word."

Mario shrugged, displaying a lack of confidence, "That won't be easy."

The commander replied with a twinkle in his eyes and a smile, "He is Etta's first cousin."

For the first time in months, Mario sensed some hope and realized that there was a glimpse of light at the end of the tunnel.

The Mario campaign started the next day. His closest friend joked about signing. Benny said, "I will finally get an opportunity to raise my hedges properly." Others declared that they could now sleep on Sunday mornings. The bocce players could play their games without interruption. Wild ducks and rabbits would get a chance to multiply. Patrick knew they were engaging in playful conversation. Mario's fun days were over since his arrest. His love

Second Chance

for hunting ceased. He never rode his horse again and never sang again.

Patrick and his team were successful. There was not a single refusal.

The parish priest, who was Mario's nemesis, willingly signed and predicted that Uncle Sam would curtail his usage of his four-letter words. Patrick knew otherwise and smiled to himself: Fat chance, not in the service. In two weeks, they had amassed a total of nearly four hundred signatures.

The big day had arrived. The commander informed Mario that they had an appointment with the Attorney General the following morning. Carmela wasn't overjoyed with the proceedings. She sat with Rocco one night and expressed her feelings. "Mario is being a good boy. He works for Mr. Morgan and spends his spare time reading. Why can't everything stay the way it is?"

Rocco replied without hesitation, "Because he's got to do what he must do."

"What kind of crazy talk are you giving me? We have one son and everybody is signing a paper to get rid of him by sending him off to war."

"Carmela, mia, you don't understand. He got himself into trouble and this is the only opportunity he has to pull himself clean with society."

"But Rocco, I don't care about society. My Mario could get killed and be buried in some far-away land. Look what happened to Filomena's son in the South Pacific. They haven't even found that poor boy's body."

Rocco shrugged as he puffed on a cigar, "Don't ever worry about Mario. Nothing bad will happen to him."

She was looking for some semblance of assurance, "How do you know?"

He looked directly into her eyes and hugged her, "Carmela, I had a dream about Mario and it was about Italy. He visited our relatives and came home safely."

She smiled briefly, "You don't believe in dreams."

"I never have, but I believe in this one." Carmela believed very strongly in dreams. Rocco conveniently fabricated this one.

His wife made one last statement, "If anything bad happens to my Mario, I will find a way to leave your life forever. God is my witness."

Pico

The next morning, there was no delaying. Patrick Morgan was dressed in a dark blue suit and resembled a successful executive. Mario arrived as directed, dressed in a sport's jacket and tie. Etta kissed Patrick as she did every morning and on this day, she heartily hugged and kissed Mario. Etta had tears in her eyes as she warmly pushed him away, "Good luck, Mario Calcagni." Patrick in the car was getting impatient.

"Thank you, Etta."

The commander pointed a finger foward, "To the State House."

Mario was quiet during the ride and thought to himself: What have I got to lose? If I'm denied the chance, I'll get a second job and save my money with the goal of starting my own business. If I get the opportunity, it's what I really want.

Patrick didn't want to tip his hand for obvious reasons. Unknown to Mario and everyone else with the exception of Etta and the Attorney General, Mario was accepted. Patrick wanted to dramatize the significance of the meeting and the Attorney General insisted on meeting Mario.

They entered the massive, white public building and walked along the lengthy hall. Their shoes echoed a tapping cadence on the marble floor. Mario was apprehensive. Patrick pressed his arm with assurance as Mario sighed. They entered the outer office and the secretary greeted them. She smiled and opened the inner office door as the Attorney General rose and came foward to greet Patrick.

He introduced himself to Mario and told them to be seated. He was a short, middle-aged Irishman who looked and acted like James Cagney. The Attorney General came right to the point, "So, you're the young man who wants to serve our country."

"Yes, sir," replied Mario.

He clasped his hands behind the back of his neck as he leaned back in his chair, "Son, a lot of people have a great deal of faith in you."

Mario didn't answer, but nodded in the affirmative.

"I have never witnessed such a display of positive reaction where so many fine people supported an individual, especially in the case of Mr. Morgan here."

"I understand, sir, and I appreciate their efforts regardless of the outcome of this meeting."

The Attorney General rose from his chair and approached Mario, "Son, Patrick convinced me a few months ago. I made my decision at that time. Your paperwork was processed a few days ago."

Mario was speechless and could only smile with happiness. "Mario, this is the first time that I have done this. Make me proud. In fact make all your backers proud of you."

Mario looked at Patrick then back at the Attorney General, "Sir, rest assured, you made the right decision."

The Attorney General walked a few feet with his hand on Mario's elbow, "Tell me, son, did you really beat the hell out of Johnny Catalano?" Mario smiled in the affirmative. "You don't need to answer that question but I know for a fact that a lot of law enforcement officers are elated. He has been brought down a few pegs."

As they departed, they offered their thanks. The Attorney General made one final parting remark, "Mario, someday, if you ever decide to enter politics, do it as a Democrat."

Mario was walking at a faster pace and raised his thumb with an agreeable salute. They neared the car and Mario commented, "Great guy with class, but one thing bothered me."

"What could possibly bother you about that man?"

"It's nothing, but when he talked to me, he didn't look me in the eye."

"I guess you don't know."

"Don't know what?"

"The Attorney General is blind and has been since he was a child."

Mario reiterated with awe, "Classy guy." He turned to Morgan, his eyes filled with tears, "Talk about class, the category was produced for one Patrick Morgan."

Within two weeks, Mario Calcagni was in the United States Army. During Christmas, it was learned that Mario was on the Cassino front. Patrick commented to a group of friends one Sunday after Mass, "By God, I feel sorry for whatever enemy comes into contact with Mario."

6
Peaceful Pico

SANTA LUCCICIA

Sul Mare Lucia.

L'Astro D' Argento

Placida E L'Onda

Prospero II Vento

Venite All' Agile Barchetta Mia

Santa Lucia, Santa Lucia

O Dolce Napoli

O Sol Beato

Dove Sorridere Volle II Creato

Tu Sei L' Impero Del Firmamento

Santa Lucia, Santa Lucia

    (From An Old Italian Song)

Peaceful Pico

Angelina was a slight girl small for her age but far beyond her peers in wisdom. She was adventurous and possessed excellent physical abilities. However, her greatest attribute was her voice. Whenever she sang, everything around her appeared to come to life. The song Santa Lucia was her favorite and she sang it often. Her cousin Gabriella sang along with her as they toiled in the fields.

The mountain peasants worked hard but were happy. There were times when Angelina's voice would echo from a nearby mountain. As she concluded one verse, another far away peasant would pick up where Angelina left off and continue along with the next verse, and still, another would continue from a farther location.

Angelina smiled gracefully and announced, "Listen to what I started. I can do that anytime that I wish." When the round-robin singing was completed, Angelina would start over and Gabriella followed.

Her mother, Anna, working in the same field, was proud of her eight year old daughter. Gabriella was six years older than her but could never keep up with the mischievous child. They were more like sisters and rightfully so. Gabriella's mother died during her childbirth and her father was killed during a battle in World War One. Gabriella was reared by Anna who was her mother's sister. Indeed, the girls were sisters and it was debatable who was older. They would race to the outer limits of the field and Angelina always won. There were times when Angelina made-up sad stories and moved Gabriella to crying. Then she would tell her that she was only kidding.

The constant singing gave the impression that everyone sang and consequently, they were all happy. For the most part, the threat of Mussolini and his fascists was commonly ignored by the villagers.

The fall harvesting brought everyone together and they helped each other. The one thing the peasants insisted on was that Angelina be present for singing and she never let them down. The little girl was the shining voice of the mountain community. She made harvesting a gala occasion. During the evening, even though most were tired, they ate, enjoyed the wine, sang and danced. The next morning, the entire program was repeated at another peasant's property. It was a seasonal community project for survival.

Every Monday morning, it was Angelina's responsibility to bake the bread for the week. Her mother would knead the dough early

Pico

in the morning and instruct her daughter. "I better not smell smoke on the bread. Remember, after the oven gets hot, remove the ashes, wipe the oven floor and then bake the bread." Angelina would look at her mother with pleading eyes and look at the heavens asking for assistance. The responsibility was difficult but Anna had confidence in Angelina.

The first time she made bread, the girl was distracted by her pet cow and forgot to remove the ashes. The mishap resulted in the bread having the odor of smoke. Anna became enraged and scolded Angelina. She ranted, "It's all your father's fault."

"Mama, Papa is in America, he didn't bake the bread."

Whenever Anna had a setback, she would reiterate, "If it wasn't for your father, I wouldn't be in this mess."

"But Mama, if it wasn't for Papa, we wouldn't have this house, barn and ten acres of decent land." Her mother looked at her with a smile and thought how fortunate she was to have Angelina. Her husband Angelo was well-acquainted with war. He fought with distinction on the Austrian front during World War I. He was wounded twice. America was now his home and he was a citizen of his adopted country. Angelo made up his mind years previously. There was no way that his sons were going to fight for a country that had illusions of recreating the Roman Empire.

Anna refused to immigrate and kept her youngest daughter with her. She felt that she had to care for her parents. The family was apart but whenever Anna encountered a problem, Angelo got the blame. She cursed him under her breath for at least five minutes. The lingering memories of six years ago were etched in her mind for life. Anna once confided in her best friend, Anita, "I would rather endure a war here than live in America with Angelo."

Sunday was the best day of the week. After attending Mass in the morning, most people visited friends and relatives. The Sunday meal was the only time that meat was served and the main staple, pasta, referred to as Sunday pasta because the combination was eggs and flour only, no water. Weekdays, the pasta was made with water added to the eggs and semolina flour.

The Italians are religious people, especially the women peasants. Nearly every town had a shrine and the women spend a great deal of their spare time visiting various shrines. Some of the men were either in the military service or working in cities or other

Peaceful Pico

countries. The trips to shrines were festive occasions. A luncheon was packed and the participants formed a procession.

Once a year, they walked to Itri for Madonna Della Civita. The legend has it that a very large tree fell approximately four feet from the ground. An altar was built on the stump and designated as a shrine. The particular saint associated with this location healed the deaf. San Giovanni had a shrine. Likewise Pastena, San Giorgo, Ciprano, etc.

There were religious occasions for every time of the year. Some were more important than others. Most notedly, the feast of Santo Guiseppi and the feast of Santo Antonio.

Between work, religion, the feast days and shrine trips, the local peasantry were fully scheduled. However, they were begining to feel the demands of an ambitious dictator. Life had been happy but change was looming on the mountainous horizon.

Government officials were becoming a problem. They started by taxing copper utensils and Anna feared, "What could be next?" As usual, she cursed her husband in America. Young boys now were marching proudly and subsequently introduced to Mussolini doctrines. His bravado impressed many who lived in cities and small towns but Anna saw Mussolini as womanizer who made complete fools of peasant women. She did not approve of such men.

Anna was concerned over a rumor that Mussolini was about to confiscate gold jewelry. He demanded that the gold was required for the war effort. Like most women, she was adverse to relinquishing her gold.

Angelina noticed her mother's concern. "Mama, why are you so worried?"

"The best thing that your father did was giving me some gold jewelry. Every time he came back from America, he always gave me a valuable gift of gold. Now, we are going to lose it all, everything."

Angelina said firmly, "Don't give them anything. Tell the officials that you are poor and don't have anything. Give them a loaf of bread."

Gabriella got into the discussion, "Give them a bottle of grandfather's wine."

"You girls are young and don't understand. There are jealous people around who will tell them what I have. I could go to jail and you girls will be alone."

Pico

Angelina recommended, "Tell them that you lost the ring while working and someone stole the necklace from you in Naples."

Gabriella said, "It could sooner be true. Luigi Calcagni said there are many thieves in Naples."

Anna took Angelina and Gabriella in her arms, "You are brave children. The three of us are a family and we are alone. If your father was a proper family man and he was here, I would give them nothing."

"Mama, we are not alone, we have grandfather, grandmother, your half-brothers and half-sisters."

"Don't worry about it children, I'll think about something," reassured Anna.

During the night, Angelina thought about the gold problem. The girl reasoned that her mother was too poor to give away her assets. The next morning she feigned she was too ill to work in the fields. Gabriella admonished Angelina, "You're not sick." Angelina glared at her.

Anna looked over her daughter and believed her because she never lied. Before leaving, she instructed her. "If you feel better later, gather some wood and prepare a fire for this evening."

"Yes, Mama."

As soon as they left, Angelina went through her mother's personal belongings and selected all the gold jewelry. The only piece of gold missing was the wedding ring on Anna's finger. She wrapped the items carefully and tightly in a leather pouch. Angelina dug a small deep hole in the earthen kitchen floor, buried the pouch and covered the hole. Angelina pounded the soft earth with a heavy stone to conceal the valuables. Angelina was positive that she was helping her mother keep her gold assets. Evening approached as she gathered some loose twigs and branches for a fire. When Anna and Gabriella arrived, the house was warm and comfortable.

The girl knew that as soon as she informed Anna that the gold was safely hidden, Angelo would begin receiving every curse known to mankind. "Mama, I did something today and you might be very angry."

"What kind of trouble did you get into now?"

"I took all the gold and buried it in a safe place."

Anna looked at her adoringly, "Come here, I want to embrace you."

Peaceful Pico

Angelina said, "I thought you would be angry and start cursing the father that I've never seen."

"I'll tell you why I'm not angry. I intended to do the same thing tonight. You beat me to it. When they come, I'll give them my wedding ring. Your father will not be around, so I will act as if I'm not married."

"Mama, if Papa ever came back, would you love him?"

Anna hugged her little girl. "Don't be concerned with adult problems. Go to bed, say your prayers and have happy dreams."

The very next day, the government officials came. A few dogs followed barking at them. One of the officials, Pietro, a short and overweight man, kicked one of the dogs. Angelina scolded, "You shouldn't be cruel to animals."

He replied, "Keep these dogs away from me or I'll shoot them."

Anna embraced Angelina and spoke with deception, "These are very important people, show them some respect."

The official who appeared to be in charge brushed the dust off his jacket and informed Anna, "You have a duty to hand over whatever gold in your posession for the glory of Italy. After Mussolini accomplishes his goal to re-establish the Roman Empire, you will be compensated accordingly."

"What happens if he fails?"

The official testily replied, "El Duce will succeed because he is a man of destiny."

"Well, you came to the wrong place, I have no gold but you are welcome to have a cup of coffee."

"Signora Micheletti, don't be coy with us, we know that your husband gave you gold as gifts."

"Yes, many years ago but we are poor peasants and traded some with a gypsy for a mule. I went to Naples once and was robbed."

The official asked arrogantly, "You mean to stand there and tell me that you have no gold?"

"I have only the ring that's on my finger."

"Then you must give it to, us but if you're withholding any more gold, you will be arrested."

She smirked as she struggled to remove her ring, "I hope you're satisfied, it's the only possession I have." Anna gave them the wedding ring along with a string of curses. She was aggressively profane and they were happy to leave her property.

Pico

Angelina's grandfather, Antonio, had a fertile five-acre grape orchard. He was reputed to produce the best wine in the entire province. Antonio had a generous heart but was a serious individual. He would not tolerate incompetence or laziness and many people feared him. He was physically strong and mentally aware of his surroundings. Antonio was a no-nonsense man, who was honest and highly respected by the few individuals who dared make his acquaintance.

Whenever he saw the beautiful Angelina, he would flash a broad, loving smile. He never smiled unless Angelina was present. The girl could do no wrong. He was completely oblivious to her mischief. They often walked together and inspected the grapes. "Angelina, taste the grape. Is it ripe enough to make the wine?" He smiled as the child selected a grape and placed it in her mouth. She rolled it around for a few moments and absorbed the taste, "Yes, Grandfather," she announced happily, "it is ready!" He laughed heartily and lifted her on his shoulders. "Someday, you will become a great wine maker."

Angelina sat on a stone wall observing her grandfather testing the grape. He walked away and entered his wine cave to retrieve a bottle of wine with a glass. He gave a sigh of relief as he sat next to his granddaughter and poured a glass, "Here, take a little sip." She went through the motions learned by observation, looking at the wine and smelled the aroma. She took a sip and emulated him with an approving sigh. Antonio smiled and asked, "What is my little angel thinking about?"

"Nonnino, I was reading a book about Caesar Augustus and came across something I don't understand."

He was well acquainted with Roman history and asked, "What is it about the great Augustus that you don't understand?"

"I can't figure it out, why did he and his people add water with their wine?"

"That's a very good question and an interesting point, I can recall reading the same thing." He winked, "I don't know the reason. What do you think?"

The girl thought for awhile and then replied, "I'm sure that the Emperor had all the wine he desired so he wasn't about to conserve the wine by diluting it."

"Augustus hosted many parties and each one was a festival."

"Then the reason must be that they wanted to enjoy themselves for a long period without getting drunk."

He hugged her with pride, "I believe that you answered your own question. You solved one question, what else is on your mind?"

"Papa, do you think the war will ever come to Italy?"

"No, my little angel, don't despair. The war will never come here."

"Why?"

"Because the regional mountains are on our side. They will protect us. Commanders of enemy armies would be stupid to even think of an invasion. This terrain is easy to defend, difficult to assault."

"I hope you're right because I think that I shall be afraid of war."

"There won't be any war here, trust me." Antonio relaxed and swallowed some wine, he was comfortable but knew that Angelina wasn't. "Tell me, what else is bothering you?"

The little girl looked at her grandfather, "I love you, Nonnino. You are a good man."

"Thank you, little one. I don't dislike anyone. I just want to be alone whenever I so desire. There are times that I can't communicate with anyone so instead of making a fool of myself, I remain quiet."

Angelina became pensive. Antonio sensed that she was leading up to something else but hesitated. "Nipotina, tell me the real reason for this conversation, which I'm enjoying more that you can imagine."

"Mama mentioned that people have told her that you've been sad since the day that her mother died. Did you love my grandmother?"

Tears formed in Antonio's eyes, no one has ever been close enough to him to ask about his life of long ago. When Anna was born, her mother was only nineteen years old when she died during childbirth. Antonio was never the same. He took Angelina's hands, "You resemble your grandmother in every way. You sing like her, you run like her, you're exactly the same. I knew her since she was a child. You have all her qualities. She was a gift from God."

"Mama said you loved her very much."

"Angelina, I know there wasn't a greater love between two people. God gave her to me and then took her away. In later years,

Pico

I remarried but no one could replace my beloved Gracia. I still think of her every day. That is why I'm quiet."

"Nonnino, I understand and it will always be our secret."

He hugged her, "I hope you experience the same love that your grandmother and I had."

"I love you, Grandfather."

"I love you even more, Nipotina."

The wine festival season was approaching. Most of the local grape growers would keep a close eye on Antonio as he was their barometer. When he began harvesting the grapes, they knew it was time. They kept their distance from him but agreed that he was a master wine-maker. As the days of the grape picking grew closer, the growers of the area placed an observer on a nearby hill. Antonio was under constant daylight surveillance.

On the morning that he began harvesting, the observer informed others. "Antonio has begun." The news spread like wildfire. In a way, Antonio decided the commencement of the Wine Festival.

Italy is a land of art, music, food, religion and wine. The combination resulted in a series of celebrations. The peasants found a way to make work a happy occasion.

Another tradition evolved during the wine festival season. It became the prime time for young lovers to elope. During the celebration of days following honoring saints, parents kept close watch on lovers. However, the wine festival was a different matter. The wine flowed and the adults overindulged. They sang, danced and sipped. Consequently, security laxed and the young lovers found the opportunity to disappear into the hills for romantic interludes. Needless to say, the young couples would elope to a neighboring village and marry. These incidents were resorted to because the same parents didn't condone the match up by lovers whose parents disapproved of the relationships.

Antonio raised a glass of wine during one evening meal and proclaimed, "Tomorrow, we will begin the picking the grape." Everyone in the family became excited. The work force consisted of Antonio, his wife, two sons, two daughters, granddaughters Angelina and Gabriella. In addition, many close relatives participated. He added to no one in particular, "Be sure to inform Luigi Calcagni. Between him, Angelina and Gabriella, the singing

Peaceful Pico

will melodiously resound off the mountains. The saints will be happy and bless us with excellent wine."

They began at sunrise. The women wore kerchiefs over their heads and wraparound peasant dresses. Each carried a basket that was soon filled with grapes. They sang as they harvested. Antonio would stop, remove his hat and wipe the sweat from his forehead. He marveled at Angelina. She not only had the most beautiful voice but she was also the fastest picker. The workers deposited the full baskets into a large container that resembled the bottom section of a vat.

Luigi arrived with his accordion. Angelina heard him singing on the trail and she answered him with her high soprano voice. She guided him in with a version of Santa Lucia.

Antonio soon signaled that it was time to begin crushing the grapes and the fun started. Anna, Angelina, Gabriella, along with other female relatives, washed their feet and jumped into the vat. Luigi began playing as the women sang and stomped on the grapes. The men continued to deposit additional baskets as the tireless women continued with their revelry. The adults drank wine from previous bottling. The children were allowed small portions.

Antonio was a loner, but he loved his family. He ruled his clan with a firm hand, but was a fair man. Today was his day, he was happy and occasionally joined in with the singing. However, it was imperative to harvest, crush the grapes and initiate the fermentation on the first day. Antonio believed that the fruit must not be allowed to stand. Freshness to the fermentation was the criterion. Previous years, he had dug a cave into the side of a hill on his property with a pick and shovel. He conveyed, "It must be kept cool from the first fermentation to final ageing."

The large cave held twenty oak barrels that were placed upright position twenty inches from the floor and the top section removed. The crushed grapes were poured into the barrels to within twelve inches from the top and covered with protective toppings. The cover was utilized as a protection against the fruit mosquitoes. Antonio was aware that the purpose of the blanket covers was to assist in promoting fermentation. However, he had another reason. Antonio suspected that the fruit flies induced wine to turn to vinegar. He felt that some wines had a very slight trace of a vinegary taste and deduced that some wine makers were careless about the fruit flies.

Pico

Antonio made the best wine because he perfected every phase of the fermentation process, including the elimination of the fruit flies. Antonio handed down the secrets to his immediate family but took greater time in educating Angelina. He would say, "Remember, Angelina, don't divulge our secret of wine making to anyone."

She reassured the old man, "Don't worry, Grandfather, I won't."

The sun soon set and they built a warming fire. They roasted sausages along with peppers. Cheese and crusty bread accompanied the meal. They drank more wine. The singing and dancing continued late into the night. The celebrations occurred in larger neighboring locales, but on a similar scale.

The next day, Anna told the girls, "Today we are going up the mountains to gather wood. Then tomorrow morning, we will go to Pontecorvo to sell it."

Angelina offered, "Mama, why can't we go down the hill to Pico and sell it there.?"

"Because, little one, pottery makers in Pontecorvo need wood and we will make more money there."

"But, Mama, it's a long walk there and will take a lot of time. Isn't it wiser to pick more wood and sell it for less money close by?"

"Angelina, why can't you be quiet like Gabriella?"

"Mama, we can make three or four trips to nearby Pico and make more money."

Anna sighed, "Have you forgotten about Guido, the forest ranger? He is like a hawk and will fine us as soon as we pick wood in an illegal area."

"No, he won't, as long as Gabriella is with us. He is in love with her."

Anna looked at Gabriella, "I don't want you to have anything to do with him."

"Don't worry, Mama, I hate that man." Gabriella added, "We heard that a few days ago, he gave a ticket to Loretta, the gypsy, for picking up a few sticks along a path."

Anna revealed with delight, "That is why we are going to pick wood in the forbidden sector. Loretta is going to give him the evil eyes tonight and he will have an unbearable headache tomorrow."

Angelina laughed, "The woods will be full of people because Guido won't be there. That's why, Mama, we should forget about the distant Pontecorvo."

Peaceful Pico

"You know, Angelina, you may be right, thanks to the efforts of Loretta the gypsy. We must take advantage of this opportunity because Guido never gets sick."

As predicted, Guido had the evil eyes and the mountain peasants took advantage of the situation. They collected enough wood to sell both in Pico and Pontecorvo. They needed the money because Italy now showed signs of joining Hitler's war.

Life continued as usual. The work, singing, processions and festivals proceeded except that some young men were missing as they were called up into the service. The old and very young males assisted the hard working peasant women. The old men sang with enthusiasm and the young boys worked for the first time. The government officials soon confiscated all the pots and pans made of copper and iron. Every thing was now to be cooked with clay pottery which, ironically, produced better results. However, a change in lifestyle loomed on the horizon.

7

The Little Big Lady

Mussolini joined the war with Hitler. He was worried that Germany would dominate Europe and Italy would be left behind. He had numerous reasons to stay out of the war. However, outside of Mussolini and his Fascists, Italians were not concerned with the war. The multitudes of the population had no interest in the new Roman empire. Another reason was that Italy had no resource that were essential for war. The country was never ready for a conflict. An additional mistake was that he had the opportunity to join the Allies but greed prevailed and he jumped on the Hitler juggernaut.

History has recorded that all his decisions were doomed to fail. After countless setbacks, King Victor Emanuel, had Mussolini arrested and imprisoned. Italy surrendered and most of the country, except for the hard-core Fascists who continued with the Axis.

The event triggered a mass exodus of Italian soldiers to return to their villages and cities. There was no problem in southern Italy because most of the territory was occupied by Allied forces. In fact, the United States authorized 300,000 Italian soldiers to bear arms. However, in German occupied zones, the conditions were totally different. The soldiers who returned to these areas threw their weapons in the nearest ravine and contemplated joining their families.

Giovanni and Luigi Calcagni were two brothers who abandoned the war and returned to Pico. They were eager to rejoin their families and tried to spend as much time with them as they dared.

The Little Big Lady

They, along with other men in their same situation, hid in the mountains. However, Giovanni was extremely close to his mother who was ill in health. He was the youngest of the children and was accused of being her favorite.

Giovanni had somehow procured a hen that produced a fresh egg nearly every day. The egg was important because Giovanni beat it in a cup of wine and made his mother drink it. The problem was that she wouldn't cooperate unless Giovanni administered the concoction and she began regaining her health. He, however, was in jeopardy. Every time he entered Pico, his chances of being apprehended escalated.

One morning, as the hen laid an egg with a loud cackle, a German soldier was nearby. There were few chickens left because the occupying troops confiscated them. Everyone knew the story behind this particular hen. The Picanos were aware that it belonged to the former Captain Giovanni Calcagni who distinguished himself on the Russian front. He received the Iron Cross for heroism.

When Italy surrendered, his unit became prisoners of war. As they were being escorted back to a rear area, he and a fellow Picano jumped a German guard and shot their way to freedom. The local troops were alerted but they didn't know what he looked like.

His wife, Anita, warned him to stay out of the village. However, he was fearless and was dedicated to his mother's welfare. Anita pleaded, "Your mother is fine, you will die before her. Giovanni, you have a wife and three children."

"After what I've been through, do you think I'm worried about these rear guard asinos?"

She shook her head, it's no use talking to him, he never listened to anyone. The Picanos praised him and looked upon Giovanni as their leader. They agreed with Anita and urged him to spend more time in the mountains.

The inevitable happened. A German discovered the hen and confiscated it. Giovanni slipped into town to perform his daily task. The chicken was gone. Giovanni went into a rage and cursed every saint he could think of.

Anita said, "I'm happy it's finally over. Now you can pay more attention to your safety."

"My mother will surely perish." He sat in a chair and buried his head in his hands.

Pico

"Your mother will be fine, think about me and the children." He paid no heed to Anita.

"Mama," he cried out, "how can I continue to care for you?"

His mother heard him from her little bedroom and answered, "Don't worry, my Giovanni, I'll survive."

He didn't want to hear this. "Which German took the chicken? I'll kill the son of a whore."

Anita replied, "Don't ask me, how should I know?"

"Where were you? Didn't I tell you to keep an eye on that chicken?"

"My dearest Giovanni, I'm out trying to buy or forage food for the children. I can't stay here all day long babysitting a chicken."

Giovanni threw his arms in the air in desperation, "I was going to bring the chicken into the mountains but I always figured that the cackle would reveal our position." He was leaving sooner than expected.

"Where are you going?"

He reached the door and then returned. Giovanni spoke softly, "Can't you see? He killed my mother." He turned and left abruptly.

"Giovanni," she shouted, "Your mother will be fine."

Pietro, who lived next door heard the conversation. He met her near the front door, "Is your husband really going to kill a German over a chicken?"

"Yes, and soon all hell will break loose."

Pietro shook his head and twirled his index finger while pointing to his head and looking skyward in amazement. Anita said sadly, "Yes, Giovanni is crazy and has no fear." Pietro said, "Perhaps it was the Russian front, I heard it was very bad there. Maybe he will cool off or find another hen."

"There are no more chickens around, the Germans took them all. I know Giovanni too well. He will retaliate, all he can think about now is that his mother could die."

"Perhaps his brother Luigi can talk some sense into him."

"Are you serious? Luigi is just as bad as him. I blame their father. When he died the kids became very independent."

"Yes," said Pietro. "They turned out fine, everybody respects them."

Anita said, "Not after Giovanni kills a German as a revenge over a chicken."

The Little Big Lady

Giovanni went down into the ravine where he had previously discarded his weapon. He searched for a pistol, a weapon that could be hidden. The terrain was rocky and with a great deal of overgrowth. He searched for nearly an hour and finally found his own pistol. Giovanni went back to his hiding place, a small cave in the upper mountain areas. He cleaned the weapon and reloaded it. Giovanni had an idea where to look for the chicken stealer because an old woman tipped him on the location. She had seen the German with a chicken and Giovanni was sure that there was only one chicken in town. He planned to carry out his plan alone. Giovanni knew the mountains and low-lying hills like the back of his hand. In two weeks, he knew every German position.

Darkness arrived and Giovanni headed for the location indicated by the old lady. He was still fuming and cursing to himself as he spotted three soldiers around a small fire. They were drinking wine and laughing at their jokes.

Giovanni crawled behind a large boulder. He could hear them and knew they couldn't see him against the glare of the bright fire. They were roasting a bird on a spit. Giovanni wanted to be sure that it was his chicken and not some wild bird. He knew enough German language to understand the conversation. The moon was bright and he had a clear shot. Off in the distant mountains, he could hear the rumble of artillery bombardment. He thought to himself: I have a terrible feeling about this area. The Germans were fortifying defensive positions near Pontecorvo. Giovanni kept listening with patience. As one turned the bird on the spit, he said, "Those Italians fed this hen more than they feed themselves." Another joked, "They won't have to feed this one anymore." Giovanni was prone as he aimed his pistol at them. He decided to shoot as they were eating.

Meanwhile, the thundering noise from the Cassino area continued. It seemed to have little effect with the revelers. He wondered: What the hell are they doing here besides drinking and eating? Perhaps they're off-duty. His patience was wearing off, he decided to move.

He was approximately seventy paces away. He leveled his pistol and took aim. "This is for my mother," as he fired off four quick shots. One fell to the ground. The other two were wounded and managed to leave the fire site. Giovanni leaped foward and began

Pico

chasing the two. One struggled to a nearby tree for support. Giovanni caught up with him and shot the soldier in the head. The third eluded him and made it back to his headquarters.

Giovanni made his way back to his cave. Soon sirens began blaring in the night. The garrison squad was all over Pico rounding up every civilian in sight. They kept them in the village square all night. The next morning, they began rounding up everyone in the hills. The rumor was that they would be killed by a firing squad.

Anita, with her mother-in-law and three children, managed to elude the Germans. She took them to her friend Anna's farmhouse out of Pico. Not only were Angelina and Gabriella there but some fifty people from the village. Anna couldn't turn them out. They decided, if they kept quiet, there was a chance. She cursed under her breath, "All this trouble over a goddamn chicken."

People, fearing summary execution, cried, "We are all going to be shot." Apparently, no one knew that Giovanni was responsible for their perilous situation. The only person who knew was Pietro and he wouldn't tell anyone. He loved him like a son.

Two German soldiers came on the property and stopped near the well for a drink. Angelina was observing them through a crack in the door. She was relaying every move they made. Anita took command, "Quickly, Anna, get me some food and wine."

Angelina watched and said, "They are looking this way and are preparing to come here."

Anita had the food with wine, she turned to Anna, "Wish me luck. How do I look?"

Anna smiled briefly, "Good luck and you look fine."

Angelina opened the door as Anita stepped outdoors. In addition to the wine, she had goat cheese with black olives and bread. Anita approached the soldiers and noticed that they were no more than youngsters. Anita was one of the most beautiful women in Pico. She walked with an alluring gait that caught their attention.

Angelina was still peeking through the crack and said, "If Giovanni saw her walking like that, he would break her legs."

Anita smiled broadly and spoke to them in their native tongue, "You men look tired and hungry. Can I share my food and wine with you?" They hesitated, but her smile, beauty and knowledge of the German language overcame their sense of duty. They looked around and were charmed by the serenity of the small farm. One

The Little Big Lady

of the soldiers said, "Oh, what the heck." The others followed suit. "We will be happy to share your wine and food. Understand, we must deliver the people of this area to the village square."

Anita ignored the last remark. She patiently described the proper way to enjoy the lunch. "You must have a clean white tablecloth," as she fluttered and placed it over the widest section of the stone well. Then we lay out the meal, and with a smile she asked, "Doesn't it all look appealing?" The Teduscis agreed enthusiastically. The combination of black olives and amber-colored cheese, brown crusty bread and red wine on a white table cloth was indeed inviting. They were ready to dig in when she courteously said to them, "No, no," she laughed pleasantly, "there is a proper sequence in enjoying the meal to its fullest."

Anita poured wine into three glasses, "Watch me carefully and follow my every step." First she took an olive, removed the pit and replaced it in her mouth. "Take a bite of cheese along with the bread. It is important to chew all three at the same time. The most important thing is a sip of wine to mix with the food. There is nothing like it."

They followed her procedures. Their facial expressions displayed satisfaction. One of the boys commented, "I believe that your presence makes the meal a success." Anita pinched his cheek and poured them both more wine, "You are a charmer." The other boy said, "We, Germans, eat too much meat."

Angelina was observing this through the crack and relayed her observations, "They are all laughing and having a good time. It appears that Anita is doing well. One of the soldiers is letting her hold his rifle."

Anita asked, "Why are you herding the farmers down to the village?" She poured more wine from the large flask.

"We are not supposed to tell, but we understand that a large number will be executed."

She screamed in disbelief, "Mama mia, how can you kill me and my children?"

The blonde soldier replied, "The partisans ambushed and killed two German soldiers and wounded another. The rule is that ten civilians will be killed for every German soldier that dies from the hands of a civilian." The companion added, "That means twenty will face the firing squad. If the wounded one dies and his chances are slim, then thirty will be added to the new count."

Pico

Anita began to weep, "Me and my three small children are the only people in the house." They kept on eating without answering. "My husband fought on the Russian front. After Italy surrendered, he joined the Germans." Anita tried to take away the food. "You people are disgusting. My husband is risking his life for your Furher and you want to kill his family. Hitler will make you answer to what you are about to do because he awarded my husband the Iron Cross."

Angelina reported anew, "I think we are in trouble. Anita is crying and gathering what is left of the meal."

The blonde soldier looked passively at Anita and spoke, "Why didn't you tell us sooner about this fellow? He is our comrade-in-arms. Go to your children and be quiet."

She smiled and made the sign of the cross, "God bless you both and I hope that you will make it home safely."

"Leave the food here."

She complied and gave them each a kiss on the cheek. They blushed as beautiful Anita walked steadily back to the house. Her fingers were crossed in front of her.

One of the soldiers advised, "Be very quiet or we will all be in trouble."

She waved with approval.

Angelina said, "Here she comes," and opened the door.

Anita entered the farmhouse and was as white as a ghost. She was successful as far as the occupants in the house were concerned but the village was in danger.

Anna asked, "What's wrong? You look sick."

"The Germans have brought all the people into the square under the pretense that they require a head-count. They haven't told anyone but they are going to pick out thirty unfortunate ones for the firing squad." Antonio, along with the others, was concerned and asked, "Why?"

Anita began to cry and between sobs explained, "Because two Germans were killed and that husband of mine is the culprit."

Anna comforted her, "Giovanni must have had a good reason."

Between the sobbing, Anita laughed nervously. "Yes, he had a good reason all right; they stole his chicken and ate it."

Antonio defended Giovanni, who was one of the few people he admired. Anita was disappointed and explained, "Don't you people

The Little Big Lady

realize that thirty people are going to die?" Some of the older ladies began to say the Rosary. Anna said, "Let's think of something fast."

There were no ideas except praying. Angelina reported that the soldiers have left. There was something about the praying that lingered in Anita's mind. It was a message that she couldn't tie together. Suddenly the mental block cleared. Anita announced, "We have a chance to save them! The Commandant is a Roman Catholic. We must get word to the priest so he can intervene."

Someone in the crowd suggested, "Perhaps he knows."

"No, nobody knows. They suspect something but have no idea of the fate that awaits them. We need a volunteer to inform Father Notte."

Anita felt responsible because Giovanni was the cause and volunteered, "I'll go."

Anna replied, "No, you won't, you have three children."

Angelina stepped foward, "I am the only person here who can get the message to the priest."

Her grandfather answered sternly, "No, you won't, not my Angelina."

The girl stood tall and made a case for herself. "I can run faster than anyone here. I know the trails, even the ones that are blocked by rock slides. There are no Germans along the way that I'll take. Besides, they won't shoot at a little girl."

Most agreed with Angelina, understandably so because they were afraid. Antonio wouldn't hear of it and said, "I'll go." His family disagreed by pointing out that the old man would need to travel the main road. It was most likely he will be apprehended and added to the group.

Angelina's bravery was only exceeded by her confidence. She embraced her grandfather and assured him that she would return safely. The old man never showed a weakness but at that moment, he had tears in his eyes and retired to the back of the room.

Anna, though unconvinced, had the utmost confidence in her daughter. The girl was a survivor and something very special. She had a strong feeling that nothing bad would ever happen to her Angelina. It was a time for embraces, all of them directed to the little heroine-to-be. Before departing, he mother called her back and said, "Be careful, I love you."

"Don't worry, Mama, you will know when I come back."

Pico

She departed with a wave as the old folks interrupted their Rosary by asking the Almighty to protect Angelina.

Anita told Anna, "Don't worry, she will get to the priest."

"I'm not too concerned with her contacting the priest. I worry more about the good Father stopping the Commandant."

Angelina quickly disappeared along a trail that hadn't been used in recent years except by her. In some areas, it was blocked by small rock slides from heavy rains. However, her adventurous excursions gave her command of the trail. Lately, she observed Giovanni Calcagni using sections of it. It was by no means a short-cut. In reality, it traversed an arc around a large hill that didn't quite qualify as a mountain. The end of the trail was covered with dense brush and came out to within one hundred yards of the church. Under the circumstances, she made rapid progress. Within twenty minutes, Angelina was ready to make her way into town. She looked around and momentarily thought to herself: Now I know why Giovanni used this trail. The captain was one of the few men that she admired and respected. Angelina liked all the Calcagnis.

There were perhaps ten buildings and a few streets that resembled alleys between her and the objective. It was eerily quiet. Most of the inhabitants were confined to the center of town. She made her move and quickly ran towards the connecting houses. She entered an alley and turned a corner at full-speed when she suddenly came upon a German sentry. He was caught by surprise and managed to command, "Halt."

Angelina darted by him without stopping and veered quickly to the right. By the time he reached the corner, Angelina was safely in church. With the holy water, she blessed herself. At the foot of the altar she genuflected. Angelina performed all the required rituals and located Father Notte in the rear of the church. She had her first chance to catch her breath, "Father, you must help the people in the square."

He was over sixty years old and the oldest priest in Pico. The town had five churches which was odd for such a small community. Father Notte's head resembled a halo because he had a circular bald spot on his head. Most priests in the area were also bald. He was a handsome, dignified man that, as a young man, was the subject of numerous rumors. The talk was that he once contemplated leaving the church for a beautiful woman. However, he was now beyond reproach and the most respected of all the priests in the area.

The Little Big Lady

"What is it, Angelina?"

She explained the story and decided that the priest needed some reinforcement when he encountered the Commandant. "Father, the Germans think that Italian partisans killed the soldiers."

The priest said, "There are no known partisans in this area, so, who do you think was responsible?"

Angelina feigned fear and acted like she knew but was afraid to be questioned by the Germans.

Father placed his arms on Angelina's shoulders, "Tell me, what do you know?"

She shied away. Father Notte assured her, "It's like confession, I won't turn you in for anything in this world." Angelina knew that the priest couldn't lie so she would lie for him. The situation dictated anything to save the people. Besides, it will be only a little white lie. "The Americans did it."

"There are no Americans in the area."

"I saw it happen. An American bomber was shot down near Pastena and I saw two parachutes in the night sky."

"Did you see the Americans on the ground?"

"Yes, Father, they asked me for directions. One of them spoke Italian. I didn't see them shoot the Germans but they were headed in the direction where the event took place."

Father Notte made the sign of the cross and instructed her to stay behind. "Father, I'm coming with you," she begged.

"No, my child. It is too dangerous for you."

"Perhaps, I can help."

The priest looked at her and realized her determination, "Okay, stay close to me and be quiet."

They approached the village square, hand-in-hand, the priest and the determined little Angelina. Together, they would do verbal battle with the Commandant of the German Command.

They entered the square as the crowd displayed a sense of relief. The priest has arrived and was sure to get the people released. Thirty men and women were lined up against a building wall. The women were weeping. The firing squad was in formation. Father Notte and Angelina walked towards the Commandant. The German leader said, "I am glad that you are here, Father. I want you to administer the last rites to these unfortunates."

The priest asked, "Why are you committing the sin of killing? They are not enemy soldiers, these are innocent civilians."

The Little Big Lady

"Father, we have a rule. If civilians kill our soldiers, we retaliate by a ratio of ten-to-one."

Angelina spoke, "The Americans shot the soldiers. You can't catch them so you're looking for a way out."

The Commandant turned to his aide. "Who the hell is this little girl?"

She stared him in the eyes. "My name is Angelina and I'm not afraid of you." The German leader was a good Catholic and was looking for a way out so he wouldn't be obliged to shoot the civilians. The appearance of the girl was a stroke of luck for the Commandant. The priest stretched the truth, "I heard the same story that American flyers were shot down and they committed the act."

The Commandant never wanted to kill innocent civilians and asked Angelina, "Did you see the Americans?"

"Yes, I saw them and I later heard shooting."

He deliberated momentarily. The decision was made. The truth didn't matter, and he thought: Thank God for this brave little girl, she enabled me to avoid a catastrophy. The crowd was tense. The priest was silent as Angelina stared bravely at the German leader. He kept a stern face and shouted "Avanti! Avanti before I change my mind." The line consisted mostly of women. They all thanked the Commandant from a distance. The priest looked him in the eye and said, "There is a place for you in heaven. God bless you."

Everyone dispersed but Angelina who remained standing in front of him.

He bowed to her and said in perfect Italian, "Thank you, little Angelina. Thank you for coming here today."

She turned with a satisfied smile on her face and ran off to the hills. The German uttered, "There goes a little big lady."

The people in the farmhouse were expecting the worse. It was a long time since Angelina left. Her grandfather was irritated with everyone. He knew it was a mistake to send a child out to do a man's job. Anna was ready to burst into tears. She had a handkerchief in her hand and continuously wiped the tears. Gabriella was emulating Anna. Anita felt terrible and cursed Giovanni. The old ladies were still praying with the rosary beads. Everyone was downhearted. The house resembled a morgue.

The Little Big Lady

Suddenly, Anna perked up. She thought she heard a familiar noise in the distance. The young Gabriella screamed with joy, "It's Angelina! It's Angelina!" The song became louder. Angelina was singing her favorite Santa Lucia.

On this day, her voice was challenging the echoes themselves. They opened the door and all went out to meet Angelina. They could hear her but not see her. She was too far away. After a few moments, Angelina turned a curve in the lane and was visible for the first time. They couldn't believe their eyes. Half the village accompanied her home. They made their presence known by singing along with Angelina.

The Commandant wasn't about to be caught again in the position he recently experienced. Fighting for the Furher and being of the Roman Catholic faith was in deep conflict to each other. He decided to stabilize the situation by dispersing a percentage of his command into the local mountain farm areas. He directed his soldiers to set up small encampments on many farms.

8

Horst and the Michelettis

It was early fall when Angelina was on a fig tree, picking the tasty fruit. She looked out and saw four German soldiers approaching on horseback. Their uniforms were immaculate and their riding boots had a mirrored luster. The horses were spirited and well-groomed. The men were good riders and handled the steeds with confidence. She saw troops before in the hills but never on horseback. Gabriella called Angelina to come down from the tree. Anna came out of the house as the men dismounted. The soldier in charge was a Corporal. He looked at Gabriella with a smile. He turned to Anna and said in Italian, "We have been assigned to encamp on your property. We are going to use your barn for the horses. Other than that, you can go about your business and we will not bother you. In essence, we are here to protect you from the enemy."

Angelina said, "Your enemy has been in the same place for a long time, they never come here."

Corp. Horst kept his smile on Gabriella as he spoke, "I hope the enemy never comes here. We will pitch a tent and be self-sustaining. We cook our own food and wash our own clothes."

"Will you protect us from those parasitic Italian government officials?"

"Sure, why not, if they bother you, I'll take care of them." Anna and her daughters were elated, it was too good to be true. Corp. Horst said, "There could be a curfew at a later date, but if that happens, it would be for your protection."

Horst and the Michelettis

The following days revealed that the soldiers were friendly. They shared their rations with the peasants. Within a short time they received about twenty more horses and located some of them in the barn. The soldiers built a corral for the rest of the horses. They intended to care for the animals. In the morning they would let the animals run freely in the corral, and in the afternoon, wash and groom them. When a horse was troublesome, Horst would tie the animal to a tree and whip it. The mistreatment of the animal didn't bother Angelina, but it did upset Gabriella.

One day, she approached Horst at one of his training sessions, "Why are you beating that beautiful horse?" It was the first time that she spoke to Horst. He addressed her many times but she avoided him.

"Your sister Angelina told me that it bothers you when I train a horse."

"Angelina is not my sister, she's is my cousin."

"I hear you call her mother, Mama."

Gabriella picked a flower from the ground. "Anna is my aunt and she raised me, so she is my mother and yes, Angelina is my sister. You never answered my question as to why you whip the horse."

Horst gave her a pleasant smile and said, "Gabriella, you are the most beautiful girl I have ever seen." She blushed. "I mean it with all my heart."

She continued to smile as she picked another flower. "But why do you beat that horse.?"

"Because I knew that it was the only way I could draw some attention from you."

"Horst, you are playing games with me."

He laughed, "Because you are so beautiful, I will tell you the truth."

Gabriella sat down on the grass and spread her dress out around her. "I'm all ears."

Horst was quiet as he picked flowers and handed them to Gabriella. Anna observed them from the farmhouse. Corporal Horst yelled, "Good morning, Mama." She waved and began hanging the wash on the clothesline. "This stallion is the most spirited of our horses and it belongs to the Commandant and I don't particularly care for officers."

"Well," emphasized a bewildered Gabriella, "you better be careful because Angelina knows the Commandant."

Pico

Horst replied, "Do you mean that our little Angelina here is the famous Angelina who brought the Commandant down to size?"

"Yes, that's my sister."

"She is your sister and then again she's your cousin. I'm confused. Gabriella, can I visit with you and your family tonight?"

She didn't hesitate. "Yes you may, but I must first ask permission from Mama."

"How will I know?"

"I will go now and ask her. If she approves, I shall raise my hand upwards."

"And what if she disapproves?" Gabriella laughed as they both rose. He assisted her by holding her hands. "Horst, you must be a dummy. It stands to reason, I won't raise my hand at all." She left to join Anna.

"By the way, the Commandant knows this spirited horse and he has directed me to whip him into order. I'm only following orders."

Gabriella reached the farmhouse and spoke with Anna. She turned to face the handsome German and decided to tease him a little. She watched him for a minute as he became impatient and waved his hand furiously. When she reasoned that he had had enough, Gabriella raised her hand as high as she could.

They sat near the fireplace and he did a great deal of talking. Like many of the Germans, he was fluent with Italian. He spoke mostly of his family. They lived on a small farm in the foothills of Bavaria. He had a younger brother and older sister. They farmed grain for making beer. They also had a few horses and domestic farm animals. Horst said he was an excellent skier and had a chance to join an elite Army mountain group until he hurt his leg.

Gabriella asked, "Does your leg bother you?"

"No, that was a long time ago, I'm fine now. How are you all getting along?" Anna and Gabriella shrugged as Angelina lay sleeping on the floor. Horst continued, "War is bad. No one really wins. Soldiers are wounded or die on both sides. Civilians suffer the most."

Anna said, "Food is a big problem. Some of the German and Czechoslovakian soldiers confiscated most of our corn and chickens. Other than that, we are surviving."

Horst was dismayed to hear about the looting. He promised that he would do his best to supply them with food. Anna believed him because he appeared to be warm and sincere.

Horst and the Michelettis

Horst thanked them for a wonderful evening as Gabriella walked him to the door. They stepped outside. The campsite was some two hundred feet from the farmhouse. Soldiers were singing "Lili Marlene" accompanied by a guitar and sounded happy.

Horst said, "Do you know that German song?"

Gabriella challenged, "It's not Teudesci, it's Italian."

He shook his head, "Gabriella, you are beautiful and funny." He held her hand and released it tenderly, "Good night, lovely Gabriella, I must join my friends."

The days progressed and love became apparent between the beautiful peasant girl Gabriella and the corporal. They had much in common but most of all, they were both farmers. He was handsome, compassionate and sincere. She was the perfect farmer's daughter. Gabriella had the qualities for true love. She would always be dedicated to the man she loved. She was shy which changed dramatically when she became comfortable with people. With Horst, it was love at first sight and love became stronger as the days passed.

Angelina was the fly in the ointment. She kept a constant watch on them. Anna smiled to herself and realized: Angelina is the perfect chaperone. Horst was a breath of spring air. His company and joyful antics kept their minds off the sufferings of war. His leisure time was with Gabriella. Horst was an accepted nuisance. Gabriella planted garlic and he wound dig it up and replant it upside down. When Gabriella caught him in the act, she threw her arms up and playfully plead to the Madonna. Angelina would sit on a nearby wall and toss pellets at them.

Horst mocked her displeasure and patiently advised her, "My Gabriella, you may know about garlic but you certainly don't know how to plant the cloves. The sprout must go down before coming up."

She sat down. "I believe that you are crazy. All the Teduscis understand is beer, sausage and greasy foods. How can you teach me about garlic?"

"I'm showing you the German way. If the clove works hard to survive, the result will be strong and hardy."

"Mama will kill me if the garlic doesn't grow." She buried her head in her hands, "Please, Horst, let's do it the Italian way. Perhaps it grows that way in Bavaria but not here."

Pico

Horst replied, "Okay, I'll replant the garlic your way under one condition."

"Now, what?"

"You must sing 'Lili Marlene' for me while I replant the garlic."

"Yes, I will sing it and Angelina will help me."

Gabriella began the first chorus and Horst stopped her, "You must sing it in German."

Angelina protested, "It's an Italian song and I won't sing it in German. Besides, I don't know the German words."

"No problem, I'll teach you. Now, repeat after me." He sang the first line and she repeated in Italian.

"Gabriella, you are making fun of me."

"The words are too harsh, my version has true melody." He tried again and she defied him again and sang in Italian.

"Gabriella, you are not holding up your end of the bargain."

She playfully threw clogs of dirt at him. Horst grabbed her arms and kissed her tenderly, "Gabriella, let's get married by an Italian priest."

She looked at him seriously, "You know it's not allowed, your Army won't allow it."

"I don't care about the Army. I'll desert and hide in the mountains until the war is over."

"Horst, now I know that you are crazy."

"I'm serious, we can take your sister and mother along. After the front lines pass, we can return."

Angelina agreed, "I think it's a good idea."

Gabriella turned to her, "Mind your own business."

Horst embraced her. "The lines can't hold forever. I fear bad times for this area. It will come from both the Germans and the Allies."

Gabriella shrugged in disbelief, "Go, go inspect your men."

9
La Fattura

Guido Bandini was the local Forest Ranger who issued fines to the peasants who gathered firewood in restricted areas. The law was ridiculous because the farmers needed wood to cook their meals and keep warm during the winter months.

The authorized zones were in the upper reaches of the mountains and in most cases, required five to six hours for the round trip. Guido was not a favorite citizen of the community because he prosecuted the needy and took bribes from the wealthy. He once issued a ticket to his paternal grandfather. Guido had no mercy for his fellow man unless they gave his some compensation. Whenever he wrote out a ticket, his standard statement was: The law is the law and I uphold it to the fullest. The old man pleaded, "Why can't you look the other way? After all, I'm your grandfather?"

Guido laughed. "I'm a Fascist and I'm loyal to the great One. The new emperor is creating an empire and I will be an important part of it."

"You are mentally sick, grandson, Mussolini has already destroyed our beautiful and peaceful Italia."

"What the hell do you know, old man?" Guido retorted.

"I'll tell you what I know, you and your kind will arrive at an ugly ending and the sooner, the better."

"Is that a fact? Well, starting next week, you and the rest of the lazy asinos will be required to save all the animal manure."

The old man replied, "Why? We need it for the meager crops we are trying to cultivate."

Pico

"Too bad, Il Duce needs the manure for explosives."

His grandfather walked away in disgust but couldn't resist a parting shot, "You should mix your shit with the manure; that way, the bomb will explode with additional force."

Guido reminded the old man, "Don't forget, the fine is due in two weeks." He put his thumbs on his suspenders and extended them outward. "You old bastard, if you don't pay the fine, I'll turn you in to the Germans. Slave labor, building bunkers, will suit you to a tee."

Guido lived with his mother and sister. The only person who loved him was his mother. They shared the same mold. Guido hated his sister Nina and the feeling was mutual. She was the opposite of her mother and brother. Nina was friendly with Gabriella and Angelina and made an effort to avoid her own family as much as possible.

The fact remained, Guido was the most despised creature of the area.

His mother was constantly pressuring him to take a wife. "After all, my son, you have a prestigious position in the government. You are a fine, handsome young man." A son always appears handsome to his mother and Guido had the misfortune of believing her. He had taken his mother's advice and for the past year, had centered his eyes on the beautiful Gabriella. It would never become a reality.

She detested him more than any man. Whenever Gabriella went to Sunday mass, she prayed to be forgiven for hating Guido and asked that he be banished from her life. He tried to court her but she ignored him. Gabriella was inclined not to hurt a person's feelings and would plead to Angelina for help.

Whenever Angelina gathered wood, she would have Gabriella carry a double load because Guido would never give her a ticket. Angelina one day suggested, "I'll be up in the fig tree with a big rock. You lure him into position and I'll take care of the rest."

"Angelina, you are so mischievous. Can't you think of a better idea? Isn't there someone that can discourage him?"

"Grandfather will straighten him out, but he may wind up killing the creep."

"No, I don't want him dead because I will never be able to go to church again."

La Fattura

Angelina was sitting on the tree branch and offered, "The Calcagni brothers will do you the favor but they've disappeared into the mountains. I can imagine Giovanni giving him a backhand."

Gabriella spoke with concern, "I'm afraid of both him and his mother. I wish he would find someone else to bother."

"No way, sister, they are confident and secure."

Gabriella responded, "I know it's a sin but I will kill myself rather than yield to that despicable Guido Bandini."

"Don't talk so foolishly. He is only a man and a weak one at that."

Guido was naive into believing that Gabriella was shy and sooner or later, everything would fall into place. He could certainly afford a wife and provide a good living. He also believed that being a Fascist was an advantage for the future. Guido was determined to pursue Gabriella when Horst came on to the scene and all but destroyed what chances he thought was real. Horst was her knight in shining armor. Gabriella was relieved and elated. However, she didn't consider the jealousy fermenting in the sick mind of the forest ranger. He hated the German but there was nothing he could do about it. Guido stayed away from Gabriella. He was angry.

One night, he was sitting alone with his mother, he confided, "The German took my girl. If I can't have her, no one will."

"My poor son, what are you going to do?"

"I'll do what I must. I will shoot her in the woods and the Czechoslovakians will be blamed."

"My son, you are approaching the problem like an asino."

"Mama, what can I do? Every time I see her and that Tedusci, my insides burn."

"I have the solution. You can make her suffer until the German goes away. He can't stay here forever. She will be so frail and sick that he won't look at her anymore. After he goes, you will have the power to make her well again and she will be indebted to you. I should have thought of the idea sooner. Gabriella will marry you and only you."

His face took an inquisitive expression, "What is your idea, Mama?"

"Leave it all to me. I'll go see my cousin and he will straighten everything out." The old lady congratulated herself. "Yes, I should have thought of it sooner. Gabriella has no family to speak of. Her father is dead and she has no brothers."

Pico

"Mama, are you referring to Santo, the Fattuciare?" She replied with a smile. He was concerned for her safety. "I don't want to hurt my Gabriella in any way."

"A few minutes ago, you wanted to kill her. You are as indecisive as your father was, I never wanted you to pick up his traits. My handsome and strong son, you must have some courage and not be so lame."

Guido listened as his mother outlined her plan. "La Fattura is a hex that your uncle will administer. She will become very sick but won't die. Don't worry about anything. The Fattuciare can control the severity of her health and make her well whenever you decide. My son, you will have the power of God."

Guido had heard many tales of these witch doctors. Santo did most of his voodoo in the Itri area and hardly ever took orders around Pico. He advocated that he had morals and wouldn't practice in his own backyard.

It was generally known that the Facttuciares had hundreds of variations of poisons, herbs and spices. They also had an antidote for every conceivable brew. However, they never revealed that segment of the business. The objective was to project mystical powers during a ritual. There was no hex; it was the imagination of uneducated and naive observers.

The potion of poison was real and the antidote was equally real. The objective was to induce the potion into the victim. The method varied but there had to be some contact. The process was known as La Fattura. The knowledge of the primitive and illegal act of inducing La Fattura was handed down by family members. Consequently, the modern day Fattuciare were the family descendants of the ages. Outsiders were not permitted to learn the trade. There was no question regarding their capabilities. Now the unsuspecting Gabriella was at the mercy of Guido and his mother.

The small group that made up the society of degenerates earned considerable money. First, they were paid to make people sick and at the conclusion of the suffering, they were paid to heal the person. They had a substantial following because of the nature of the human mind. There will always be occasions where formal education is lacking. The strong and vindictive will always take advantage of the weak and innocent. Guido had no scruples. He agreed with his mother and she visited her cousin Santo.

La Fattura

She explained her intentions and Santo wanted to know the severity of sickness that Guido required. The old lady wanted Gabriella to become temporarily crippled and uncomfortably ill.

The Fattuciare warned her, "We must be careful during these times. The people have very little in the way of food and there is much malnutrition."

She replied unconcerned, "There is nothing wrong with her, the German gives her ample food."

"I'll work out something but when you want to cure her you must employ another fattuciare."

She asked in surprise, "Why can't you heal her when I ask you to?"

"My dear cousin, of course I can but I'm not permitted. We fattuciares have an agreement between ourselves."

"What kind of nonsense are you giving me?"

He explained, "If one Fattuciare administers the Fattura then another must take it away. It's a matter of economics, but even more so, it's connected to a religious belief we adhere to."

She laughed, mocking him, "People like you don't believe in religion."

She came from the same mold however, and believed his nonsense. He commanded her as she departed, "Come back tomorrow and I will have the potion ready. Remember, you are to tell no one or I will put a hex on you."

"Hell you will, I won't come to within ten feet of you."

He laughed, "Don't forget to bring the money."

"What money, I thought you were doing the family a favor?"

"My dear cousin, business is business and I have a big overhead."

The next day, she picked up the potion and paid him. He instructed, "You must use it within five days or it will become weak and ineffective."

"How do I use it?"

"You need to put it in her food and be sure that Nina knows nothing about it. In fact, you shouldn't even tell that idiot son of yours."

"I already told him, he is the one who needs the Fattura."

"Tell him that he must be silent. Men have a habit of boasting to demonstrate powers that they don't have."

Pico

The old lady knew that Gabriella trusted Nina and she intended on using the friendship to administer the poison. She had previously questioned Nina on Gabriella's habits. There was one day a week that Gabriella stayed home to do the washing, cleaning and prepare the evening meal. It was also the day that Nina visited Gabriella during the afternoon. The old lady made a dish of Minestra which consisted of boiled seasoned dandelion over crusty bread. She was in a jovial mood as she and Nina sat and enjoyed a midday lunch.

As Nina rose to leave, her mother said convincingly, "We have extra minestra, so bring some to your friend Gabriella." As Nina went to her room to get a shawl, the old lady took a small portion of minestra and added the poison potion. She wrapped a white napkin over the small round bowl and told Nina not to eat any because there was only enough for one.

"Don't worry Mama, I'm full up to here," as she placed her hand near her chin.

Nina arrived at Gabriella's house. Horst was away for over a week. She was alone and getting ready to do the housework. Gabriella asked, "What do you have there?"

"My mother made some extra minestra and sent you some for lunch."

"Why?"

"I don't know, she's been in a good mood lately, since she had the argument with Guido. She has been good to me. Have some Gabriella, it's really delicious."

Poor Nina had no idea as to what would transpire in the ensuing weeks. Gabriella ate the food and that evening became violently ill. Anna and Angelina came back from the fields and could not believe Gabriella's condition. She was on the floor, doubled up in pain and crying. Anna spent the entire night by her side but could not determine the problem.

The next day, Gabriella was paralyzed from the waist down. She was sick to the stomach and had recurring fever. Days passed and there was no improvement. Anna gave her tea, concocted from wild mountain vegetation, to no avail. Others brought their family remedies as Gabriella's condition became worse. They tried not to leave her alone as there were occasions of pain and loneliness. Whenever the fever subsided, she tried to play with the cat. The pet avoided her and she cried.

La Fattura

Anna didn't have any money, but she and Angelina sat Gabriella on the donkey and escorted her to Pico to visit the village doctor. The doctor examined her perfunctorily and informed them that there was nothing he could find. He referred them to a doctor in Pontecorvo. A few days later, they again mounted Gabriella on the donkey and traveled to Pontecorvo. The results were the same, the doctor had no idea what was wrong with her. Anna cried on the way home and cursed Angelo in America. She told Angelina, "We are helpless and your father is probably sleeping in a warm bed with some putana."

"Mama, please don't cry. If it wasn't for the war, Papa would have helped."

"We need money to help Gabriella get well. She needs a specialist. Perhaps we should visit his brother in Roccasecca and borrow money from him."

"Mama, it's wartime, nobody has money except the Fascists."

It was raining when they arrived at the farmhouse. Horst was waiting. Nina had briefed him about Gabriella. Angelina ran and embraced him, "I'm so happy that you're here. Perhaps you are the medicine that Gabriella needs." Anna hugged him and cried, "My poor, innocent Gabriella, we have tried everything and nothing helps."

Horst cradled Gabriella in his arms and tried to comfort her. She was embarrassed with her condition and wouldn't look into his face. He carried her into the house and cuddled her face against his. Her eyes were closed and she didn't reveal any emotion.

"Don't worry, my little flower, I will get our German doctor to make you well."

Anna spoke sadly, "Your German doctor will never have the time to examine her. Besides, he can probably get himself into trouble administering to an Italian peasant."

"Don't worry, Mama. I take care of his horse here and he will do me a favor."

Angelina shouted happily, "Gabriella, the German doctor is better than any of the horse doctors that are here. I know he can help you."

Gabriella had already lost over twenty pounds and was losing the will to live. Horst went to Gabriella and began assuring her that everything was going to be well. He mentioned the doctor and she screamed, "Go away, I don't want to see you or anyone else."

Pico

He looked at her with an agony that he couldn't describe. Her eyes were closed as she turned her head to the wall and began an uncontrollable sobbing. He looked at both Anna and Angelina with his palms spread upwards, "I love her so, what can I do?" The soldier was powerless. Anna tried to console him. He had tears in his eyes. She knew that Horst loved Gabriella. It wasn't a case of German meeting Italian. It wasn't a passing fancy created by the conditions of war. Anna knew that somehow, the two were destined to be together and God would provide the solution.

Two days later, the doctor paid a visit to check on his horse. Horst had the animal in peak condition as he led the animal out into the nearby field. Anna and Angelina observed from a distance as Horst confered with him. After a few minutes, the doctor went to his vehicle and fetched his medical bag.

They came to the farmhouse and Horst introduced him. He was a medium-sized gentleman with graying sideburns. He ordered everyone out of the farmhouse as he began the examination. Horst assured them that Gabriella was in good hands. After a half-hour, the doctor came out with a puzzled look on his face. He spoke in German to Horst. The girl had a fever that would be eliminated with some pills that he handed to Horst. He gave her a shot of penicillin to shake the infection. Now came the difficult part, "I'm sorry, Corporal, I've never seen a case like this before. She has symptoms of polio but as far as I can tell, it's not that."

"Sir, I love this girl, if we don't help her, she will die."

"Don't blame yourself, that poor child is in the hands of the Almighty."

Horst thought for a moment, "Perhaps I should pray."

The doctor hesitated a moment to seek the proper words, "Yes, I strongly recommend that." He gently put his arm on Horst's shoulder, "I'm truly sorry."

Horst tried to explain the doctor's diagnosis. Angelina pulled him aside, "My mother is going to bring her to church. The priest will make her well. She lost so much weight that we are afraid she will die."

Anna added, "I pray for her to get well every hour of the day. The house of God is the last resort."

Horst didn't answer. His tears were real as he waited momentarily to gain his composure. He regressed to when he saw

La Fattura

Gabriella running in the fields. He thought of all the joyful taunting and teasing. Horst was not overly religious, but looked up to the heavens and asked from his heart, "Please God, make that innocent child well again. I'll marry her even if she's paralyzed."

Angelina heard Horst and ran off into the hills sobbing.

Anna held her composure and informed Horst, "I will take her to every church in the district and I will pray to God and I will pray to the Blessed Mother." She kept ranting out of control, "I will pray to all the saints. I will take her to every shrine." Horst tried to console her but she kept on, "Gabriella is the daughter of my sister but she is also my daughter. I raised her and love her as much as I love Angelina."

Horst escorted her into the farmhouse, "Let's talk about it tomorrow morning."

The next morning they left early. Her grandfather Antonio rigged a large, solid basket on the donkey. Anna, Angelina and Gabriella's godmother, Anita, escorted her to the first of the many churches they were to visit. They begged for food to help, but Gabriella's conditioned worsened. Her fever was gone and the pain subsided but her weight continued to drop. They visited the shrines of the areas, but to no avail.

When they left the farmhouse in the morning, Horst met them and kissed her on the forehead. She looked at him with pleading eyes and tried to smile. "Today is the day, God will help you."

Anna was losing faith. She knew that the end was near. She could no longer cope with a futile situation. They were visiting the last church in the district. It was Father Nottes' church. They placed Gabriella on the altar, basket and all. By this time, the girl was nearly half her original weight. They recited the Rosary as Father Notte observed. When they completed the prayers, Anna faced a statue of Jesus and cried, "I can go no further. I am leaving this child here to die in this holy house of God. May she rest in peace." Anna slipped to her knees and cried, "I tried, I tried but nobody listens."

Father Notte came out to console Anna. He explained quietly, "You cannot leave her here."

"You don't understand, Father, I am leaving her here. I wash my hands, she is now in the care of the Almighty."

He escorted her into a private room. Angelina followed as Anita remained with the limp and near dead Gabriella. The priest spoke solemnly, "Gabriella has what you, people, refer to as the fattura.

Pico

Anna's eyes expanded in surprise, "The fattura? Who would do such a terrible thing to an innocent person like Gabriella?"

The priest was justifiably concerned and offered advice. "Understand, the fattura is not magic, nor is it a mystery. It's an explainable process where someone administers a poison. It must have come from a jealous person because the girl was so beautiful. Whoever it was, hired a fattuciare to make and arrange the administering of the potion. These people are idiots but dangerous. Unfortunately, the concoctions are secret that only another fattuciare can solve and consequently supply the antidote. The fattuciares impress the unsuspecting public that they are associated with God. They have all been excommunicated from the Church."

Anna regained her composure. She asked the priest, "Do you mean that my sweet Gabriella has the fattura?"

Father Notte replied, "I'm sure of it and there isn't a doctor around who can heal her."

Finally, Anna understood. So did Angelina. They both agreed that they recognized the symptoms. Time was of the essence. They need some answers fast. The three women loaded Gabriella on the donkey and made their way back home. Angelina said, "Don't worry, Gabriella, we know the problem now, everything will be fine." They arrived at the farmhouse and Horst was waiting.

He lifted her off the animal and carried her into the house. Anna and Angelina spoke at the same time, "We have some news." They explained everything about the fattura. Anna, Angelina and Anita began asking all the neighbors for information on voodoo. In the event that anyone had any knowledge, they gave guarded replies. Most of the peasants feared both the fattuciares and the people who employed their services.

Whenever Gabriella had the will and strength to talk, Angelina asked her about events that occurred on the day that she first became ill. She was too weak to remember and usually fell asleep.

Horst was becoming impatient. "Yesterday, you all told me that you had news. Well, what is the good news?"

"We think we know the cause. The next thing is to find out who did it and why."

"Let's move on it. I should inform my doctor."

Anna replied sadly, "Don't waste your time, he can't do anything."

La Fattura

The next afternoon, Angelina was sitting quietly alone on a tree branch. She was contemplating and trying to solve the dilemma. Some older boys were playing in the distance when she overheard the name Gabriella uttered. Sound, in the sparsely vegetated mountains carried far and clear. She listened carefully and also heard the name of Guido Bandini mentioned.

Angelina slipped down from the tree and ran over to the boys. She began questioning them about the fattura but they ignored her. Her intuition said they knew something. She reasoned: It had to be Guido. He was seeking revenge. Angelina looked for his sister, Nina, until she found her gathering wild edible vegetation in the fields. She asked her angrily, "What do you know about Gabriela and your brother Guido?"

Nina shrugged innocently, "I'm sorry for her. I pray for her every day. She is the closest and best friend that I've ever had." Nina began to cry. Angelina thought: She is like Gabriella, a weak girl who wouldn't hurt a fly.

Angelina persisted, "A few weeks ago, before Gabriella became ill, did you have any contact with her?"

"I must have seen her nearly every day of our lives. I'm sure I saw her on that day."

"Did your crazy brother touch her or give her any food?"

"Angelina, I'm surprised at you. Gabriella hates my brother and if you really want to know, I feel the same way. She would never tolerate him touching her and accepting food from him is out of the question."

"Nina, please think hard, we are running out of precious time. Your friend is dying. Did anyone give her food on that day?"

Nina thought for awhile, "Come to think of it, I brought her a bowl of minestra on the day she became ill. Surely you don't think that I would poison her?"

Angelina's eyes lit up, "Now we are getting somewhere. Who cooked the minestra?"

"My mother."

"Ha," accused Angelina, "she and Guido are two of a kind." Nina thought for a moment and explained, "I ate the same minestra."

The interrogation hit a snag. Nina and Gabriella ate the same food and only her sister became ill. Angelina became temporarily discouraged, but one important factor lingered. Nina's mother never sent food to Anna's family. Another point, why, when Gabriella was alone?

As Nina was leaving, she said, "Guido yes, but my mother would not be a part of any fattura."

"Was Guido home that day?"

"No. He is never home during the day."

Before Nina was out, Angelina asked, "Did you and Gabriella share the food from the same bowl?"

"No, I ate the minestra at my house."

Angelina was hanging on faint hope as she approached Nina, "Did you see your mother put the food in the bowl for Gabriella?"

"As a matter of fact, it was a rainy day and I went into another room to get my shawl."

For the first time since Gabriella's illness, Angelina broke into the biggest of smiles. She hugged the astonished Nina, "I love you," and she ran home with her important discovery.

Anna snapped, "Your sister is near death and you're acting like we are about to have a festival."

"I know who caused it and I know why Gabriella is sick, and the reason," Angelina said.

"What have you learned?"

Horst, standing in the open doorway, listened curiously.

"It was Guido and his mother. I heard some boys talking, then I questioned Nina. She is innocent," Angelina explained.

Anna was convinced that it was all true. She reminisced: Guido was in love with a girl that he could never have. Along comes Horst and the girl falls in love with him. Guido avoids the family even when they are gathering wood in restricted areas. Guido is known to be vindictive and last but most important, there is a fattuciare in the Bandini family.

Anna hugged Angelina, "Now all we need to do is employ the services of a good fattuciare to heal Gabriella."

Horst turned and went to his tent. Anna said, "He should be happy, what is wrong with him?" She continued, "There is a fattuciare in Arce who can cure her. We need to borrow some money."

Angelina did not take her eyes off the German's tent. Horst came out with a pistol strapped to his side as Angelina turned to her mother, "It's out of our hands. I can see that money will not be necessary."

La Fattura

"Angelina, what are you talking about?"

"Look, Mama." Anna came to the open door. Horst was attired and determined for battle. He walked to the corral and selected a gray mare. Horst reined the horse and didn't bother to saddle her.

Anna said, "Oh, my God, he is going to kill them."

Angelina ran to him as he mounted the horse. She protested, "I'm coming with you."

"No, Angelina, I have some business to attend to. I don't want you to be any part of it."

"Angelina, come back here," Anna screamed.

She paid no attention to her mother as she tried to hold the reins of the mare. "Horst, we don't have much time, Gabriella is dying and I'm the only one who can find Guido."

He contemplated as the horse was becoming impatient and Angelina was in danger of being trampled. Horst reached out with his arm and lifted the little girl onto the mare. He galloped at high speed. Angelina's hair fluttered in the wind. They slowed down as they entered the dense trail.

"Angelina, when we find this Guido, I don't want you to be present when I confront him."

"Why?"

"Because I am going to get rough with him and his conduct will determine the outcome."

"Horst, I don't care if you kill him. If you don't, I will."

"Angelina, you are going to be a real lady someday, but don't push it."

She directed him to another mountain trail that was clearer than the previous one. He galloped at a faster pace as Angelina wrapped her arms around his waist. They rode for nearly a half-hour at Angelina's direction. They passed a few peasants who were sharing the trail and Angelina asked them if they saw the forest ranger. They didn't see him and they were not interested in him. After the riders were out of sight, many wondered, "What is that Italian girl doing with a German soldier?"

The trail thinned out and became denser. The riders dismounted and Horst led the mare as he followed Angelina. She said, "This is the best of the restricted areas and if we don't see him soon, I know how to bring him here."

Horst questioned, "How are you going to do that? I can hardly see in front of me."

Pico

"Guido is the very best at his profession. He knows these mountains better than everyone combined. He can hear a chopping noise from the next mountain and can determine the shortest and fastest route to the noise. That is all that he is good for." She found a sturdy stick and walked to a large tree.

Horst asked curiously, "What are you going to do with that stick, kill a forest ranger?"

"No, dummy." She began creating a chopping noise by hitting away at the tree. "You just wait and see, Guido will see lira bills in his eyes."

Horst looked at her curiously and decided to waste fifteen minutes or so. "All right, Angelina, you can play your game while we give the horse a rest."

"You and the horse should stay out of sight. Guido has caught us many times in this area. Believe me, he will come."

Horst tied the mare to a bush behind a large rock. He sat on the ground next to the rock which gave him a view of Angelina.

Within minutes, Guido came pacing from the opposite side of the trail. She had previously gathered some branches of wood. He looked at the stack and then assumed the authority of an important government official. "I caught you again, when are you peasants going to learn to obey the law?" She stared at him with mock fear but did not reply.

He squared his to a different position, "You are alone. I hope Gabriella learns her lesson."

"What lesson? What are you talking about?"

"You know exactly what I'm referring to. She did me wrong and I got even with her."

"Gabriella never did any wrong to anyone. She is a perfectly innocent girl."

"Is that a fact? She left an important person like me for a Tedusci. It's a disgrace!"

"Guido, you are a sick person. She was never interested in you. The only person who puts up with you is your mother and that is because she carried you for nine months."

Before he could address her again, Horst struck him from behind and tossed him against a tree trunk. The beady-eyed, balding ranger was petrified. Horst gripped his thin neck with his big left hand and lifted him against the tree. Guido's legs were dangling

La Fattura

as Horst ordered Angelina to leave. She walked down the trail, but doubled back and hid close by. Horst slapped him repeatedly, "Who is your fattuciare?"

Guido couldn't answer even if he tried because the tight grip on his neck prevented him from talking. Horst released his grip and allowed him to fall to the ground. He asked him again impatiently, "Who is the fattuciare?"

Guido replied hoarsely and guardedly, "I don't know what you're talking about."

Horst picked him up by the shoulders and snagged his jacket to a branch of the tree. Guido was again dangling and Horst slapped him a few more times only this time with greater force. "Talk, you imitation of a weasel," as he continued to slap him. Guido was weakening. Horst backed off because he didn't want to beat him unconscious. Horst pulled out his pistol and placed the barrel between Guido's' beady eyes. "My friend, do you want your brains decorating this tree?"

Guido paled and trembled as he tried to put the words together. Talk was very difficult. He was looking at death in the eye.

"Please let me go. I'm a Fascist, we are on the same side."

"You will be free of problems as soon as you tell me what I want to know."

The pistol barrel grew bigger than ever, "His name is Santo, he's my mother's cousin."

Horst relaxed but he kept the pistol in place. "One more thing, where can I find him?"

"He lives about a twenty-minute walk from here near the road to Itri. Angelina knows the location of the house."

Unknown to Horst, Angelina was standing behind him. She wanted Horst to hit him a few more times.

Horst said, "I promised you that I will free you of your problems. This is for what you did to Gabriella." Before Guido had a chance to comprehend this, Horst pulled the trigger and Guido's' brains decorated the tree. He turned and nearly bumped into Angelina. "I thought I told you to stay out of sight."

She shrugged in defiance, "Let's go get Santo, I know where he lives."

They both mounted the mare and Angelina directed him to a trail that led to the Itri road.

Pico

Horst looked straight ahead and explained with fatherly advice, "It isn't proper for a young girl to witness what just transpired. You saw me murder one of your countrymen. In one way, I feel like I evened a score and in another way, I feel bad about that."

"Horst, you talk too much. You didn't murder anyone. You just killed the enemy."

He thought for a moment, "Angelina, you have a convincing way of understanding things. Yes, he was the enemy." She felt relieved because Horst was so justified.

He thought further, "Perhaps what you saw may help you in the future."

She had her arms wrapped around his waist, "How?"

"The worse part of the war will come through this stretch of mountains and, God forbid, you may experience worst." Horst tapped the mare on the side of the neck. "Angelina, how old are you?"

"I'm thirteen years old."

"God willing, you're going to be my sister-in-law someday. You look younger but have the sense of a mature young woman." Angelina smiled but didn't answer. "I am going to kill more of the enemy but we must get the antidote first."

"Yes, Horst, you are so smart."

As they neared the farmhouse Santo was outdoors feeding his mangy dog. His wife was washing clothes as the dog started barking at the riders. Santo was alerted and at the sight of Angelina, he knew the reason for the visit. There was nothing he could do but play out the string. The riders dismounted and Horst went into action.

He fired a shot over the head of the old lady and that sent her disappearing into the wooded area.

Santo ran into the house and tried to close the door behind him. Horst first hit the attacking dog with his fist and then proceeded to kick in the farmhouse door. Horst gripped Santo with the same move that he used on Guido. This time, he dispensed with the slapping and shoved the barrel of the pistol into the mouth of the fattuciare. "I have no time to waste with idle talk, Gabriella is near death and I want the antidote or I will splatter your brains all over this room." Santo's eyes bulged with fear but like Guido, he couldn't speak because of the grip on his throat. Horst continued

La Fattura

as Santo listened. "I'm going to give you a chance. I don't want to hear the slightest excuses about your stupid voodoo. Everything is cut and dry. The antidote is for your life. Do you understand me?"

Santo had peed in his pants. He motioned with a hand gesture that he understood. Horst released him. Santo went into a closet room that resembled a dirty makeshift laboratory. Horst and Angelina followed.

Horst observed as the fattuciare mixed a concoction and poured it into a small bottle. He tried to carry on a conversation with muttering commands over the potion but Horst warned him to shut up. Angelina was also observing the proceedings.

Santo protested that the girl was now learning his secrets and would expose the nature of his business to the public. Horst informed him, "Old man, your days of business as a voodoo man is over."

He handed the bottle to Horst and appeared to gain a false bravado of security, "Who are you to tell me what to do?"

Horst paid no attention and he took a nearby cup and poured a small amount of the potion. He handed it to Santo and demanded, "Drink."

Santo drank and immediately felt secure. The German would go away happy. He recommended, "Give the sick girl a tablespoon three times a day. She will be on her way to recovery within a week."

Horst handed the bottle to Angelina and in one swift motion gripped Santo by the throat and shoved the pistol barrel into his mouth. "You want to know who I am! I'll tell you who. I'm your enemy which means that you are my enemy." Santo's eyes bulged, he didn't dare move. "We are at war, you and I. You will never harm innocent people again." Horst fired and the shot sounded like a cannon in the small work shop. They mounted the horse and headed back to the farm and Gabriella.

Angelina said, "Too bad you don't have more time. You could clear out all the fattuciares in the province."

"Don't worry about it, sister-in-law, the word will soon get around and they will phase out on their own."

"I don't know, Horst, as soon as you're gone, they will get their nerve back."

Pico

"Rest assured, they will get the message. I killed one of them and they will fear that someone else will repeat what I did. They are not to be feared. They are not invincible."

"Angelina, you are so quiet. I am doing all the talking. What are you thinking about?"

"I'm only thinking about the war that will come to us. I wonder how many people that I know will perish."

Horst replied, "Don't think about it too deeply but definitely plan a way to avoid it as much as possible. How do you feel about what happened today?"

"It hasn't bothered me at all. You did them a favor by putting them out of their misery." Horst felt better as they rode along.

Anna nursed Gabriella with the antidote and within a week she began walking again. Anna held her on one side and Angelina supported her on the other side as they paced out further ever day. The frail Gabriella was soon making excellent progress. Horst gave her extra food rations and she began to gain weight. Her smile returned and her love for Horst became stronger than ever.

Her recovery was a diversion making Horst forget the confrontation with Guido and Santo. After all, it was war and he placed the incident away in memory.

The word circulated throughout the mountain communities that the days of the fattuciares was over. The rumor became that a Tedusci soldier was responsible. The family of the deceased maintained a low profile. The German command had more important things to think about. They were unconcerned about the death of two peasants.

10

Surviving On The Edge

Winter arrived and food became scarce. The Allies began bombing the Pico area as the Germans had a supply depot in the sector. There were incidents when the bombs missed the target and fell in the foothills where most of the little farms were. There was also a great deal of action on the Cassino front. The weather was unusually bad which added to the overall deteriorating conditions. The SS of the German Army began rounding up Italian males and transporting them to labor camps in Germany. There were many incidents where people simply disappeared. The rumor was that they were forced into slave labor along the numerous German lines of defense.

One November afternoon when the sun burst through the ever present clouds, Horst and Gabriella were sitting on a stone wall. "My dearest, our battalion has been ordered to relocate to a forward position."

Gabriella was stunned, "You said that you were here to protect us and now you are leaving?"

He put his arm around her, "I don't want to leave you but we are going to maintain a secondary position just north of Pontecorvo. The position is being set up to protect Pico. If we are successful, we will still be protecting you and your family. Don't worry." He tried to assure her, "The Allies won't get by us."

Gabriella's fears persisted. "I'm afraid. I know that I won't ever see you again." She began sobbing and looked at him helplessly.

"Gabriella, my Gabriella, I promise you, I will be back. Perhaps after the hostilities end, I will live here with you." She didn't answer

Pico

and continued sobbing. "Please Gabriella, we have been through hard times together. Try to be strong, the war will pass and life will continue. I can prove my love for you. We will all hide in the mountains. To hell with this war."

Gabriella replied between sobs, "My mother will never leave her parents behind."

"Gabriella, that is no problem, we can take them with us. I hear rumors that many of the civilians of Pico plan to hide in the mountains and let the war pass them by. I'll wear peasant clothes and grow a beard."

She forced a smile, "You're too tall to pass for a peasant. My grandfather, Antonio, will never leave his property and his wine. He is not interested in the war and believes that the Americans won't bother him."

Horst shifted and turned his head in disbelief, "You mean to tell me that we can't hide in the mountains to save ourselves because of the wine?" The wind was blowing a crisp breeze. "I don't believe you Italians. You won't leave with me because your mother won't leave and because her father won't leave his precious wine. Are you people serious?"

Gabriella ran a hand through her soft hair and shrugged hopelessly, "I'm sorry, Horst. I do love you so."

He kissed her tenderly and rose to leave, "I'll be back later this evening." He knew there was no chance to relocate this family. Antonio was headstrong and it would serve no purpose to talk to him. That evening, Horst directed his attention in trying to convince Anna. He told her that after his army group left, a contingency of SS troops were scheduled to replace his unit. He heard rumor that these people had no respect for the civilian population. The rumor was that they were to turn the population into a work force. The men to work on building fortifications. The women were to be designated as domestics and the young girls to be trained as nurses. He didn't want to alarm them too much and specified that, for the time being, it was just a rumor.

Anna spoke defiantly, "I will not leave my parents to go to America. Do you believe that I would leave them to possibly die at the exposure of war?"

"Anna, I don't think you understand, the Americans are humane. It would be a blessing if they liberated this area. The problem is that there are some bad African units under the Allied Command. They are notorious mountain soldiers and will probably be coming through this very sector. In any event, you must avoid those people."

She cupped Horsts' hand, "Don't worry, my husband is an American. My sons and daughters are Americans, too. There are soldiers in the American army who have relatives here in our village of Pico." She tried to assure him, "No harm will come to us."

"Anna, for everybody's sake, I hope the Americans liberate this area soon."

Horst knew that two American divisions were fighting in the west coast sector. Their objective, when the breakthrough occurred, was the Anzio Beachhead. The German High Command rationalized that the old Appian Way was the most logical route. He feared that the Moroccans were designated to spearhead an attack through the most difficult terrain. Horst knew that his intention to defect and live in hills had failed. He had no alternative but go along with the status quo. The German had some provisions that wouldn't last very long. He handed some food to Anna, "It's not much, but the best that I can do." He then sat with Gabriella and Angelina and explained, "If the bad soldiers come into this area, you girls must avoid them at any cost. As far as I can tell, we will be making a strong defense of this area and I may be able to visit again, but I doubt it very much. The SS will be behind us to make sure that we don't retreat."

He removed a shotgun from a sack. They recognized the weapon because it had originally belonged to Angelina's father. Horst had confiscated the shotgun when he first arrived months ago. They were surprised to see the weapon. He asked, "Does anyone here know how to use this weapon?"

Gabriella replied with a mild irritation, "I don't like guns and I'll never shoot one."

Horst was concerned, "Listen to me, this gun could possibly save your lives." Gabriella was still indifferent and wouldn't change her mind. He turned to Angelina who was paying strict attention. "How about you Angelina?"

"Just show me how to use it."

Horst directed, "Now, watch what I do." He pushed the lever to the right and the double barrel shotgun cracked open. He continued, "You put a shell in each barrel and close the gun into a locking position. Now, see this little button here?"

Pico

Angelina looked over his hand, "Yes, I see it."

"Okay, when it's down, the triggers are locked but when you push it up like this, it's ready to fire. Angelina, do you understand?"

"Yes."

Horst cracked open the weapon and removed the shells and closed the gun and handed it to Angelina. He instructed, "Crack it open." She did as ordered. "Now close it and push the button down." Angelina complied. "Try to fire it." She did but the triggers were locked. Without further instructions, she pushed the button upwards and fired both triggers. Click, click.

Horst was elated, "Very good, now if we had shells in the barrels, you would have blown a hole through the roof." Angelina practiced a few more times to get used to the weapon. Horst tried one more time to interest Gabriella, but to no avail. He escorted both girls out to the barn and they climbed up into the hay loft. He loaded the shotgun and hid it in a corner out of sight. "Now remember carefully Angelina, all you need to do is aim, push the button up and fire; one trigger at a time. The gun will give a kick but you won't feel a thing. Do you think you will be able to pull the trigger if it becomes necessary?"

"Yes, Horst, I am not the least bit afraid. I will pull the trigger." Gabriella displayed some interest but insisted that she could never fire the weapon.

They went back to the house and the family tried to make Horst feel at ease during his last night with them.

They ate polenta made from corn flour as it was a special occasion for them. They drank wine they had secretly saved. At the end of the night, their best friend bid them goodbye. He kissed Anna and Angelina. Horst embraced and kissed Gabriella for what seemed like minutes. He let her go and had tears in his eyes. The next morning, Horst and his squad left before dawn. Gabriella waved from a window but it was too dark for Horst to see her.

The ensuing winter months were devastating. The bombing in the entire area escalated. The explosions along the Gustav Line at the Cassino front became more intense. It rained continuously, the weather was windy and cold. There was little food left. In fact, the SS took what meager supplies that the peasants had saved. They had no wood for fires.

One day, Antonio was walking in the hills, foraging for food. A German patrol captured him and immediately put him into captive labor along the new fall-back point called the Hitler Line. He, along with other prisoners, dug into the side of the mountain to establish fortified bunkers.

Antonio had disappeared and the family feared the worst. The previous month, his two sons disappeared and they hadn't been seen since. Anna's stepmother moved into her house. Antonio had some olive oil and wine buried in the earthen floor of his wine cave. The SS troops had confiscated all his wine. Angelina knew the location of the hidden items. She and her mother dug these up and hid these in the old horse barn. They were barely able to survive on wild vegetation mixed with a little olive oil and cold water. There was no way they could concoct a minestra or a soup because they couldn't build a fire. Anna advised, "The planes can spot the smoke from in the chimney."

Angelina missed her grandfather. They had a strong bond with each other. Every morning, she would go off into the mountains to look for him. Anna was concerned about Angelina roaming about the mountains looking for her grandfather. She was afraid that the SS would pick her up for their labor force. However, Angelina was as familiar with the mountains as Guido and the Calcagni brothers were.

Anna had a greater concern and told her little girl, "The bombs are falling everywhere. You can be killed instantly."

"Mama, be serious, this farmhouse has a greater chance of being destroyed along with all its occupants. Believe me, I'm safer in the mountains than you are here. Whenever I return, I hope and pray that the house is still intact."

Anna conceded, "Perhaps you are right. There are times when the bombing is so intense that we don't dare go outdoors to forage for food." Anna added that if the end comes, they should all face it together.

There were many times when Angelina found wood in the forest but left it there. She thought and shrugged to herself; What's the use, we can't light a fire. It will draw attention.

Angelina defied her mother. Each day, she headed into a different direction. The little girl did not accept the possibility that her grandfather could be dead, either by starvation or cruel punishment.

Her daily commitment consisted of climbing a high hill or small mountain and yelling out from the top of her voice, "Tadone."

Pico

Angelina and Gabriella always referred to their grandfather as tadone. She repeated her calls over and over. There were times when German patrols neared her but she easily evaded them.

Angelina missed her grandfather so much that she would sit alone and cry. "Tadone, next fall, we must make the wine together. Where, oh, where are you?" Again, she cried and cursed Mussolini and this destructive war.

Angelina came home one day early in the afternoon and found Gabriella crying. "What is wrong? Did something happen to Horst?"

"No, but Nina is dead. She was killed by a bomb, her mother was also killed."

"I'm sorry, Gabriella, she was a good girl."

"They were looking for something to eat in the fields when the bomb hit them."

Angelina told her and Anna, "It's more dangerous here than in the high mountains."

A week after her grandfather disappeared, Angelina was on the southern section of a mountain facing the Pontecorvo area. From a distance, she noticed a large number of people working at what appeared to be forced labor. The girl called out loud and clearly, "Tadone." Antonio was one of the peasants digging the bunker fortifications. Slave labor consisted of toiling from dawn to night while taking much abuse from the guards. The Italians were told, "Your production yield is less than half of the normal German output." They were fed very little which made them frail and weak. Antonio was old but he was strong. He kept to himself and contemplated on escape. The old man talked to no one. Some of his fellow prisoners thought he was a mute. They were all exposed to the Allies' bombing. There was no relief from the daily toil and most of the workers were in a state of despair. The chances of seeing their families again were, at best, remote.

One day, Antonio heard the familiar voice of Angelina, his beloved granddaughter. He stopped digging and searched the mountain with anticipation that his granddaughter would appear. He heard the voice again, "Tadone."

Antonio screamed out with elation, "Angelina!" The other prisoners were startled and realized that Antonio could speak. He started climbing out of the ditch when a guard approached him and shoved him back. The old man ignored the guard and tilted his head towards the mountain. Again, he heard the voice of an angel, "Tadone."

He screamed at the top of his voice as he rose from the ditch, "Angelina!"

She recognized her grandfather's voice and yelled with joy, "Tadone, I have found you!"

The guard came up and jabbed him with a bayonet. Antonio tried to explain, "It's my Angelina, I've got to see her!"

"You're not going anywhere." The guard thrust his bayonet, drawing blood from the Italian. Antonio was enraged and with all his strength, struck the pointed end of his shovel into the stomach of the guard, lifting him high into the air. The German groaned helplessly as the blood squirted through his uniform. The old man yelled, "Angelina!" In his joy he failed to understand the seriousness of his act. Antonio had a smile on his face and probably never realized that he just killed a soldier.

Another guard fired three successive shots at Antonio. He staggered, fell backwards and slipped slowly into the ditch. Blood flowed from the wounds as his fellow peasants tried to comfort him during his final moments. Antonio's last word was "Angelina." He uttered it with a soft angelic smile.

Angelina had heard his voice and the gun shots. She fell to her knees and screamed. The child was hysterical and cried, "Tadone." It was more of a shrill than a call. A prisoner who witnessed the scene gave a sad response, "A morte."

The entire valley became silent as Angelina was on her knees crying while clutching clods of earth. "Tadone, the vines, the grape, the wine that you loved so much," she sobbed. "Why, oh, why did this all happen? It is my fault that you're gone. I wish I had never found you."

Angelina collapsed and lay on the ground crying hysterically. Her mind raced through memorable times that she shared with her grandfather. After a few minutes, she rose and gathered her composure. The brave child dried her tears with the hem of her dress as some bombs exploded nearby. Angelina was oblivious to the noise of destruction and could only think of her beloved grandfather. She brushed the dust from her dress and called loudly one more time, "Tadone." Angelina stood tall and silent for a few

Pico

minutes. Tears formed in her eyes as she endured the absence of sound.

She soon arrived home. The tears were gone but the memories would remain with her forever. She informed Anna that her father was dead. Anna and Gabriella embraced her with sadness as all three cried. They decided not to tell the old lady.

Angelina went outdoors and sat by the wall. Anna came out and sat next to her. "Mama, I feel responsible for my grandfather's death." She related the painful story and told Anna, "If I had not found him, he would still be alive."

"Don't ever blame yourself, he would never have come back. Starvation, maltreatment or the war itself would sooner or later, have claimed his life. Angelina, if he heard your voice he died happily."

"Mama, he did hear me because he answered. I hope that God understands Tadone. He was a good man."

"Don't worry, the Almighty will have the proper place for him in heaven."

"Mama, Horst is a good man. Even the commandant is a good German, but the troops that are here now are very bad. I wish the Americans would hurry and drive them back to where they came from."

Anna agreed, "It will all be over soon. By summer, everything will be fine."

"Yes, Mama, I think you are right but what happens between now and then?"

Anna placed an arm around her daughter and smiled. "We can only pray and hope for the best. You, I and Gabriella have been through so much. We are destined to survive."

"Yes, Mama," she squeezed her hand, "time will tell."

By late February, they were desperate. Grandmother was failing very fast. They resorted to eating bark from trees. Anna observed, "I never thought that we would become like animals foraging anything."

More bombs were now being dropped in the Lira Valley. The Allies were softening up and delaying the supply routes. Rumor had circulated that the command headquarters of General Von Senger was abandoning the Roccasecca sector and moving further north. This in itself was a sign that the Germans didn't expect to hold any of the previous defensive lines. There were times when

Surviving On The Edge

Angelina and Gabriella observed the nighttime explosions along the surrounding mountains and ridges of the Gustav line. The bombardments resembled fireworks.

One morning near the end of the dreadful winter, Gabriella called out to Anna, "Come quickly, something is wrong with grandmother!"

Anna feared the worse but she had learned to accept tragedy. Her stepmother was dead. She was a victim of malnutrition, the old were vulnerable. Anna didn't cry. Her pain was well inside her body. Gabriella and Angelina were sad but they knew that the end was inevitable. The grandmother was finally relieved of the dreadful daily encounters.

They took some boards from a stall in the barn and constructed a makeshift coffin. Anna placed her stepmother's rosary beads in her hands and wrapped the body in a blanket. They dug a grave on the property and buried the old lady.

Angelina found some early budding ginestra flowers and placed them on the grave site. The only thing they could offer besides the yellow flowers were prayers. Anna reminisced, "She wasn't my real mother but she raised me and I loved her very much. The old lady was a quiet, unassuming woman. She loved her husband and lived to serve him only. My father kept her close to their little farm. She only left to attend Mass and an occasional visit to the shrine. I don't believe that Mama had much of a life, but what she had was enrichment with her family."

Anna placed a few more ginestras in the center of the grave and cried sadly, "Goodbye, Mama, I know that you are in Heaven."

Spring arrived and dandelions graced the fields. Wild broccoli grew in abundance. These vegetables were nutritious and could be eaten raw. In fact, it was more nutritious raw than cooked.

There was a fear of escalated bombing that the peasants hardly ventured into the fields. They had a grim choice to either starve or be blown to kingdom come. There were occasions when there was a lull. A time came when the skies were free of planes.

Anna made a fire away from the house and boiled dandelion with rabe and mixed it with olive oil.

Those days, which were indeed rare, they had a feast.

The ill-fated month of May was approaching and a lull of activity along the Cassino battle sector signaled that something important was about to happen.

Pico

The Calcagni brothers came by the farm and asked the women to follow them into hiding in the upper sections of the mountains. Giovanni said, "We have moved many people but most of the local farmers refuse to leave. The Germans have chased everyone out of Pico."

Anna replied, "I'm one of the farmers. I won't leave. If the Teduscis chase me out, I'll return here as soon as they leave."

"Giovanni is right, let's go with him," Angelina said.

Anna deliberated as Giovanni tried to convince her, "The war is coming through here for sure. I know because the General has left Roccasecca and they won't let anyone back into Pico."

Luigi added, "Most of the soldiers from here are near Pontecorvo. It looks like they are going to defend Pico."

Gabriella asked them, "Has anyone seen Horst?" Giovanni replied, "Gabriella, do you seriously believe that I would hold meetings with Germans?"

"But Horst is a good German."

"Perhaps you're right, he did get rid of a fattuciare, but I would have done the same. The German saved me the trouble."

"Did you move your mother into hiding?" Anna asked.

"No, she is too sick to make such a rough trip. There isn't much time and I'm not sure what we are going to do with her."

Anna said, "Bring her here, we will care for her."

"Then I take it that you are not leaving."

Angelina said, "Mama, Horst wanted to leave the German Army and take us all to safety. You said you wouldn't leave your mother and father. They are both dead, what is your excuse now?"

"I have no excuse. It is just that I don't believe it's necessary. The war will sweep by here and everything will return to normal. The Americans will give us food. Who knows, perhaps my sons will be part of the liberating force."

Giovanni saw that Anna wasn't about to leave.

Anna turned to Gabriella and Angelina, "You girls, go with the men."

"It's the same old story, we won't leave our mother," Angelina said.

Giovanni and Luigi had a conference. "We decided to bring our mother up here. She is old and nobody is going to bother her providing that we have your approval."

Anna said, "Of course I approve. We will be honored to have her as a guest. We all love her very much."

"Thank you, Anna, we will bring her here within an hour."

As they were leaving, Luigi commented, "Hey, Gabriella, Horst may be an exception but the only good German is a dead one." She threw a rock at him.

The evacuation of Pico was an exercise in futility. The Germans escorted everybody out of town but there was neither control nor direction. People heading north and south were turned away and directed elsewhere. Others left for east and west but, were likewise ordered to move in other directions. The Germans themselves were confused. It appeared that they expected the Allies to come from any direction.

There was word that the Americans dropped paratroopers somewhere north of Arce. The civilians were in greater disorder than the Germans because they had no official guideline to follow. The soldiers had orders to prohibit civilians from dug-in sectors. The problem was that they were dug-in in every direction.

The confusion resulted in a situation where everybody was moving around in circles with out final destinations.

Actually, the unfortunates were trapped into continuous movement. Farmers were also evacuated from their homes and told to move elsewhere. When they questioned, "Where do we go?" The soldiers became angered and ordered, "Avanti." Ultimately, some of the aimlessly wandering civilians from the town reoccupied the farmhouses that were previously evacuated. In many cases, more than fifty people crammed in one farmhouse.

They kept quiet and didn't dare use the fireplace. The situation was unbearable. In many cases, misery seeks company. However, these people didn't search for misery; it found them.

After the Calcagni brothers left along the trail to Pico used by Angelina and Giovanni, the Germans came and knocked on the door.

"Wait, I'm coming," Anna said. They banged again as Anna reached the door, "What do you want?"

They ordered, "You must leave immediately."

"This is my property and you can all go to hell."

One soldier leveled his rifle at her as Anna maintained her position.

Pico

Gabriella screamed, "Mama, please listen to them."

The German commanded, "Avanti."

"Shoot me, I would rather die on my own property."

The soldier hesitated, "I have my orders."

Anna retorted, "To hell with your stupid orders. How would you like it if your mother was in my place and a Russian was pointing a gun at her?"

The German looked into her intense eyes and lowered his weapon. He waved off his companion and turned to leave. Without facing her he advised, "Be absolutely quiet and nobody will be here." He added, "I mean that no Germans will be here."

They were fortunate that they encountered a compassionate soldier. Anna went to the door to thank him. He stopped and looked at her sadly, "Signora, there are some bad African soldiers coming this way. They have committed some atrocities in Esperia. You and your daughters must hide and avoid them."

"Are they coming here?"

"They are mountain fighters and have broken through the line."

Anna put her hand flat against her mouth, "Oh, my God." The Germans left, they were on their way to rouse other farmers out of their homes.

Within the hour, the Calcagni brothers returned with their mother. They had hidden her in a small grove. Not surprisingly, more than twenty people followed them to the house. They were desperate and had trust in the Calcagnis. Anna called Giovanni aside and related her confrontation with the German. He shrugged in despair. "It is one of the bad things associated with war. The Americans and British are honorable soldiers but the French-Africans are another story."

Angelina came to him, "Giovanni, are you still going into the high mountains?"

He pressed the child against him as if she were his own, "Yes, even more so now that I hear what your mother has told me. You are all welcome to come along."

Anna replied, "Thank you Giovanni but I think we will be safe here."

He felt sure that his mother's advanced age would be her protection. The Calcagni brothers left with their family. Anna and her two daughters waved from the door. They wished each other luck.

11

Mario Arrives At Monte Cassino Area

It was a cold, damp January night. Private First Class Mario Calcagni had joined his combat unit as a replacement. His chances of seeing action were definite.

Mario was relieved to have completed the ocean voyage. The travel from the United States to Britain was unbearable. Soldiers had to contend with stormy weather as well as the confined conditions of troop transportation. Approximately ten thousand men were billeted below deck and subsequently were packed like cattle. Fourteen soldiers slept in an area of three hundred cubic feet! Once one got into his assigned bunk area, it was an ordeal to get out. Fresh water for showers was nonexistent. Many soldiers avoided salt water showers. Many became seasick and some puked in the billeted area. There existed a stench that was beyond description. Mario would later experience worse: the odor of decaying bodies.

The ship's galley was another problem. There was no provisions for seats so one ate in a standing, rolling position. The terrible odor discouraged most. The consensus was: why eat and puke? Fortunately, everyone was forced to spend the day on deck. The fresh air was invigorating but the rough seas curtailed normal activities. Some played poker and when some lost, watches and rings were offered. Others passed their time by playing Monopoly.

For Mario, the ocean trip was the worst experience of his life. He dreaded the day he would make the return voyage home. Mario reevaluated his last thought: Hell, when that time comes, I'll probably settle for a row boat. The last leg from Britain to Italy wasn't so bad. The sea was calmer and the ship contained less passengers. They debarked at Naples and boarded trucks heading north to the Monte Cassino area. The objective was to cross a river near the Cassino area and assist a weakly established bridgehead. The rumor was that they were being thrown into the battle to try and rescue the unfortunate soldiers who were trapped. What was left of two regiments had dug in approximately five hundred yards from the river. The rescue force faced an impossible task ahead of them. The enemy had a large number of machine guns that raked the trapped Americans with no relief in sight.

Another problem was the large amount of barbed wire directly in front of them. Even worse, was the high saturation of artillery and mortar salvos which exploded with accuracy. In summing up the situation, the trapped Americans couldn't retreat even if they wanted to. In the event that anyone was fortunate enough to fall back to the river, his chances of getting across were limited.

The Germans had a clear advantage in every phase of the battle. The observation and fire power that covered the river itself was formidable in both planning and execution.

The task facing Mario's unit, at best, was difficult. Each squad carried their boats along a swampy area to the jump-off point at the edge of the sweeping river. The enemy fire was heavy in the swamp area and many in the regiment never made it to the river. Others who reached the area were pushed off into the rapids, capsized and drowned. The entire operation was an exercise in frustration.

Sgt. Peters remarked, "The son of a bitch that planned this fiasco should be hung by his balls." A member of his group added dejectedly, "We have about as much chance as a snowball in hell."

The Lieutenant barked out orders that held an uncertain future, "Let's get these boats into the water."

Mario was anxious to get across. He was concerned with the conditions and the gunfire and mortar explosions. An up-river dam was purposely opened to flood the entire section. The condition caused another difficult obstacle. The logistics were rendered nearly useless. Very few squads made it to the river's edge. Sgt. Peter's men got into their boat and shoved off. They were successful only because some of them had previous experience with row boats.

Mario Arrives At Monte Cassino Area

The distance across was less than fifty feet and they had a great deal of difficulty in maintaining a straight course. The squad reached the opposite shore and were faced with an embankment. It served as a temporary shield from the constant fire from the enemy. The night was pitch black and illumination was provided only by the light emitted from the constant explosions.

Sgt. Peters hesitated as long as he could without revealing his state of futility. Perhaps he was stalling for time, hoping that the officers would call off the operation. The decision to move was made for him. The boat they used floated down the rapids and nearly crashed into the Lieutenant's boat as he made his way across.

Soon there was a lull in the firing, and Sgt. Peter ordered the men to follow him over the embankment. They couldn't believe the scene. Dead bodies everywhere. When the shooting commenced, some utilized the dead as a shield.

Mario crawled with his body pressed hard to the ground. His forward movement was measured in inches rather than feet and yards. He moved cautiously and heard occasional groans from the nearby wounded. Some called out for the medics but Mario assumed that the medics were also dead. He lost contact with his squad.

Mario kept inching forward. Dawn was breaking and he halted. He moved cautiously to scan his surroundings. He observed a carnage of dead bodies, all Americans. The shooting had stopped again except for an occasional round of small fire. Mario reasoned: They've got to cool off those machine guns.

He reminisced the days when he hunted in the forest and kept perfectly still as to not alert the prey. However, his present situation was entirely different. The prey could shoot back and they had more firepower than he could ever imagine.

As daylight came, he noticed movement in all directions. Sgt. Peters and a few members of his squad were to his left. He recognized others from his company on his right. The soldiers they were supposed to rescue were dug in ahead and around him. Many were wounded or dead.

Daylight was accompanied by a trace of fog. Mario wasn't sure if it was fog or smoke. The battlefield was literally strewn with an even greater number of dead American infantrymen than he previously imagined. The survivors of the two-day battle were understandably incoherent and discouraged. Their will to fight

Pico

evaporated. There were a few officers to lead them but, even they appeared to be useless. In fact, the only coherent officer around was Mario's Lieutenant who had reached Sgt. Peters position. It was obvious that they were involved in a rescue position.

Mario knew that as soon as the enemy discovered them they would be in deep shit. Hell, they would have little chance to get out alive.

Mario estimated that there were perhaps fifty American survivors who could probably make a run for it. He also guessed that his reinforcement group was no more than thirty men. There was activity nearly a hundred yards ahead of him and he knew it was an enemy position.

Mario also noticed an American machine gun emplacement some 20 yards ahead of him and to the right. He contemplated: If I can get to that position in one piece, I can do some damage. He further surmised that he had a better chance there than staying put or going back.

The emplacement had some cover and Mario gained a spark of confidence. The Lieutenant observed Mario and was relieved to notice his intentions. Reaching the position without being discovered by the Germans, he was closer to them than he realized. The American quietly moved the dead out of the way and assessed his overall position and ammunition capabilities. After noting everything was still on hand, it was obvious that the dead occupants didn't have a chance to even fire on the enemy. They were cut down as soon as they set up the emplacement. Mario knew that he was in a better position to cover the retreat of his comrades.

The Lieutenant rationalized that whatever small chance they all had to get out of there was in the hands of Mario Calcagni. He passed the word that as soon as Mario opened up, they were to make a dash for the river.

Mario sized up the enemy positions. They hadn't seen him yet. He began a calculated firing and yelled at the top of his voice, "Okay, you fucking Krauts, Mario Calcagni is here."

His initial sweep wiped out three machine gun nests. He continued a steady firing as the Lieutenant ordered, "Move out, everybody for himself." Some froze in their prone position while most ran for the river. The Germans were caught off guard as a number of the Americans reached the river. They opened up with

Mario Arrives At Monte Cassino Area

a mortar barrage and managed to inflict high casualties. Meanwhile, Mario was receiving a barrage of ground fire and mortar.

Sgt. Peters was hit and most likely dead. The Lieutenant made it back to the river. He yelled at Mario to fall back. He thought to himself: The poor devil doesn't have a chance.

Mario knew that everybody was out of there except for the few that were shell shocked. There was no way he could help them. He suddenly felt a sharp pain in his right shoulder and knew immediately that it was mortar shrapnel.

Mario ducked down in his ditch and laid out his grenades. Luckily, he was left-handed and immediately started throwing grenades wildly at the enemy positions. Some scored and some didn't but he threw so many in a short time that he caused enough havoc to allow him to make his move. The Lieutenant coaxed, "Run, Mario, run."

Mario reasoned: It's now or never, if I stay here, I'm a goner. The bullets were streaming by him. Officers and most of the survivors were in the river as the shooting continued. A few drowned and others were shot in the water.

Amazingly, Mario wasn't hit and was running fast and he dove off the bank of the river and swam deeply underwater as he knew the harmless bullets hit the surface of the water. He couldn't believe his good fortune and thanked his father for teaching him to swim.

Mario reached the other side of the river bank and found the Lieutenant behind the cover of an abandoned boat. They were, by no means, safe. The Germans were still firing mortar rounds into the swamp area. The officer had a wound in the leg but it wasn't bad. Mario was bleeding from the shoulder but didn't feel any pain. The Lieutenant was helped by a medic and asked Mario to tag along.

"It looks like you got yourself a Purple Heart," Mario bantered.

"Mario, you not only got your Purple Heart but I'm going to be sure that you receive the Distinguished Service Cross," the officer replied.

"Thank you, sir."

"No, Mario, you got it all ass backwards. I thank you for saving all those lives."

"Sir, I think you're missing the point, we got the crap kicked out of us."

Pico

"Yes, but it wasn't our fault. Someone higher really fucked-up."

"I'll tell you one thing, you're one fighting son-of-a-bitch. You didn't lose soldier, you won," the officer grimaced in pain from his wound. He could not walk any further and the medic finally got him onto a stretcher.

As he laid on the stretcher, he shook Mario's hand and informed him, "By the way, you're now Sgt. Mario Calcagni." Mario gave him a salute and headed off towards the field hospital. He was looking for Sgt. Peters and what remained of his squad and suddenly remembered that Peters was dead. In fact, only two other people in his squad survived.

Mario, downhearted, thought: You make friends from day one and you either get separated from each other or someone dies in battle. Hell, it isn't worth making friends. He kicked the dirt in front of him as he mumbled to himself: I kind of liked Sgt. Peters, but he had a bad feeling about this entire operation. The man was right but could do nothing about it.

Mario entered the medical tent. The medics removed his jacket and attended the wound which appeared to be superficial. There were many in the big tent who recognized Mario and thanked him for saving their lives.

"I can't believe that you made it. The bullets around you must have seemed like raindrops," one soldier said.

Mario smiled for the first time, "Yeah, you're right and I ran in between them."

The scuttlebutt was that the Division sustained approximately fifty percent casualties. Mario evaluated the information and thought: It should never have happened to an American Division. We have such superiority in men, resources and fire power. He shook his head in disbelief.

For the time being, the division was relocated to a rear area for the purpose of replenishing their lost manpower. The following Sunday, Mario went to Mass and thanked the Almighty for sparing his life. He vowed that if he ever returned home safely, he planned to attend church every Sunday. A few weeks later, Sgt. Mario Calcagni received the Distinguished Service Cross from the Division Commander.

The General was somewhat subdued but he managed a grateful smile for Mario. He was impressed with the Division Commander

Mario Arrives At Monte Cassino Area

and heard that he was being made the scapegoat for bridgehead failure. Mario returned a salute and sensed that the General will someday have his day of glory.

In the days ahead, the battle-hardened veterans would be busy training the new replacements. The recently-decorated sergeant worked hard with his squad. He discouraged individual performance and stressed the importance of teamwork.

"We will never abandon a wounded member of this group." He continued with reluctant authority, "I don't give a fuck how tough it's going to be in those mountains but I promise you, nobody in this squad is going to die."

He taught them hand signs that signified various operations while under fire. "In this war, the radio is our best friend. I assure you, it will save lives. If it comes to choosing between the radio and a weapon, retain the radio."

"What's it like up there?," one of his men asked.

"I'll tell you what it's like. The Krauts are excellent soldiers but we're better."

"I don't know about that," one of his men smirked.

Mario gave him a silent and penetrating look before answering. "I should knock you on your ass but I'll be a good guy and explain. Those bastards are good but they are sticklers for following orders. They are spread out in ideal fortifications but are on their own. They are disciplined but in this type of war where the adversaries are sometimes within one hundred feet of each other, their system will backfire. Now, in the event that we, Americans, get into trouble we have what is called imagination. In a given situation, we can act accordingly within the bounds of some flexibility. The Germans, as individuals, don't make decisions. In fact, there will be times when they can be conned into surrendering. Remember, the most important factor is that we are street-wise and in war, it comes handy." He looked over his squad and asked, "Do I make myself perfectly clear?" They all agreed and seemed convinced that they were indeed in good hands of the fearless sergeant.

Mario wrote home, but didn't want to worry the family. He informed them that he was enjoying Italy. "The weather is great and the Italian civilians are doing as well as expected under the circumstances." He wrote a little white lie, "We beat the hell out of the Germans." He continued, "I expect to see Pico soon, it's only

Pico

about fifteen miles from here. Oh, by the way, I was awarded the Distinguished Service Cross." Mario was unaware that the event was already published in their local newspapers. He had a chance to rest and recuperate and spent a good deal of time thinking of home. He wasn't alone, everybody thought of home. He thought: Thank God I lived near the ocean and learned to swim like a fish. It was the same thought he had at the river.

Mario recalled when he and his dad went quahauging in Narragansett Bay. They would dig for the prized shellfish with their feet and toes. Once the feet revealed the shell, he would dig it out with his hand. As the tide rose, it was necessary to go underwater to retrieve the quahaug. Mario became so adept at it that he even went at high tide. The neighbors thought he was crazy but he did it anyway.

He spent many nights looking north towards the mountains and was amused by the different colors of far explosions. It was like the Fourth of July this winter.

He wondered about his relatives in Pico: Do they have food and shelter? Are they being mistreated by the enemy? Will I get the opportunity to see them soon? He then smiled to himself and thought of Pete Colombo. A few years ago, Pete went back to Pico to retire there. Poor Pete, he thought: The old man really fucked-up. Pietro was a small-time cement contractor who could neither read nor write. However, he made up for the deficiency by working very hard. He knew his trade well but was better known as a slave driver. Pietro was a good man, well-liked and a lot of fun but when it came to work, he was all business. It was customary in the neighborhood for a father to threaten an errant teenage son, "If you don't shape up, I'm going to get you a job working for Pete Colombo." The threat usually worked. He didn't pay much, but the experience was an education.

One night, Rocco was in The Sons of Italy Club, enjoying a glass of beer with Pietro. "You know Pete, I think it's time for Mario to get a little training, can you use him?"

The old man leaned back in his chair, "Of course, I can. Is the boy in trouble?"

"No, I just want to get him out of his uncle's poolroom."

"I understand, Rocco. Tell him to be at my storage building by 7 a.m."

Mario Arrives At Monte Cassino Area

Summer vacation had started and Mario thought he was going to have fun. However, when his father informed him that he enlisted his services with Mr. Colombo, Mario didn't even flinch. The next morning, bright and early, the fifteen year old Mario began his career as a construction laborer. The odds were that hardly anyone lasted over a week. Some said that Johnny Catalano held the record for three weeks.

Pete had a unique approach he used with new, so-called employees. On the first day he would work side-by-side and demonstrate his concept of labor. Mario was no exception. He arrived at the job site and Colombo singled out Mario. "Grab a shovel and come with me. You and I are going to move that pile of sand." They each got on opposite sides of the mound and began shoveling into the bed of a truck. They were even in speed, but after ten minutes or so, Mario challenged, "Come on Mr. Colombo, let's race and see who's faster." They went at it, fast and furiously. The rest of the workers applauded Mario. Even Pete tried to hide a smile. The old man murmered, "I'll fix him." He had him carrying a hundred pound cement bags for the rest of the day. It didn't have much effect on Mario because he sang all day long.

The next day, he had the boy working alone, shoveling a mixture of sand and cement into the mixer. Mario, if anything, was enjoying himself by singing whenever he got the opportunity.

Pietro worked him as hard as he could, but Mario was actually having fun. After a week, Pietro met Rocco in the club. Again, they were enjoying a glass of beer.

Rocco asked the question that many fathers had asked in the past. "Well Pete, how's my son doing?"

The old man leaned back in his chair, this time he propped it on the two rear legs as his back touched the wall. He took a sip of beer and shook his head sideways with a smile. "You know, Rocco, I like that fucking kid." It was the greatest compliment that Pete Colombo ever gave anyone and it was considered a lifetime endorsement.

Mario not only worked that entire summer but worked for him every summer and whenever Pietro asked him. The experience not only molded Mario to appreciate hard work but also built him up physically. By the time he was seventeen years of age, he was nearly six feet tall and weighed slightly over two hundred pounds. Mario was as strong as a bull and as fast as a cat.

Pico

Mario took one last look towards the mountains before retiring to his tent and said, "Mr. Colombo, stay well and try to survive until I get there."

12

Home Town Elation

The telephone began ringing an early January morning at the Rocco Calcagni household.

It was approximately 6:30 a.m. when Benny called. "Rocco, did you see this morning's newspaper?"

"No, and what the hell are you doing calling me this early? For Christ sakes, the roosters are still asleep."

"Well, you should be happy because there is a story about Mario on the bottom section of the front page."

Rocco began stringing Benny along with tongue in cheek, "What in the hell has happened now? I thought Uncle Sam would straighten him out but I guess trouble follows my Mario around. It seems that a cloud hangs over his head."

Benny interrupted him, "Rocco, you got it all wrong, Mario pulled it off. The kid hit it 'big time.' He was awarded the Distinguished Service Cross."

"What cross are you talking about?"

"You mean to tell me that you don't know anything about it?"

Rocco was having fun, he wanted to hear about Mario's exploits from someone else. Rocco had a fatherly interest to praise his son to anyone who would listen.

"Yeah, I heard a little about it. Last week, a jerk of a newspaperman called and started to ask a lot of questions about Mario, so I told him to go fuck himself."

"Rocco, that guy was on your side, he wrote a great story about your son."

Pico

"Bullshit, it sounded to me that he was just breaking balls."

Benny became confused, "Didn't you know anything about the award?"

"Sure I knew, we received a letter from him a few weeks ago and all he said was that the weather was fine and he won some kind of an award."

"Are you serious, Rocco? He referred to the Distinguished Service Cross as only an award?"

"That's all I know, believe me, I thought he was referring to some Italian lottery. You know Mario, he loves to gamble. His crazy uncle Hogan broke him in."

"Rocco, calm down and go buy a newspaper. Mario cleaned house at a river section near Cassino."

Benny hung up the telephone and commented to his wife, "That God damn Rocco, I still don't know if he realizes the heroics performed by his son."

"Now Mario can plow through the hedges all day long," Benny's wife replied.

Benny laughed, "I always liked that kid."

The telephone rang again and it was Emilio. Bookies read the newspapers very early because they check to see the horse entries for the day.

Emilio was exhuberant, "Rocco, did you see the morning paper?"

"No, why?"

"The front page, Mario is all over the front page. The paper shows that he knocked out over a hundred Germans and in turn, allowed nearly as many trapped Americans to escape!"

Rocco finally realized what really happened and told Emilio hurriedly, "I've got to go pick up a newspaper."

He left Emilio hanging as he replaced the receiver.

"Hey, Carmela, wake up, the telephone is ringing off the hook."

"What's going on, did something happen to Mario?"

"Yes, something did happen to your son," and before she became worried, he continued, "all good."

Carmela came to the kitchen as the telephone rang again.

He ordered, "Don't answer it." Rocco explained the newspaper article to his wife. She smiled with relief. The telephone rang again. She answered it as Rocco dressed. She told the caller, "Wait a minute. Rocco, where are you going?"

Home Town Elation

"I've got to pick up a newspaper, I'll be back later."

Her friend, Anna, was on the telephone as Carmela said, "That son of a bitch." Anna said, "You better make sure he didn't hear you," as she started explaining the article. In fact, Anna read the entire story to Carmela. She added, "You should see the picture of him, he looks the part."

Rocco walked out onto the street as it began to snow lightly. Rocco knew that this weather usually led to a heavy snow storm. He was heading towards the corner store when he noticed a familiar form approaching. It was Patrick Morgan and the closer he got, the bigger his smile.

Patrick embraced Rocco and spoke with elation, "He did it, the boy did the near impossible but they should have awarded him the Medal of Honor." Rocco thought proudly for a moment: Coming from the commander, it signified the importance of Mario's accomplishments.

The big Irishman had the newspaper with him and showed Rocco, "Look at the picture, the General is decorating him." Patrick was cold, his fingers trembled as he was pointing out the highlights of the story.

Rocco said, "Let's go to the Sons of Italy club for a shot of cognac. But first, I've got to buy a paper and some cigars at the corner store."

Patrick looked at his watch and said, "It's too early, they don't open until 11 a.m."

Rocco smirked arrogantly, "Don't worry about it. Tony, with the wooden leg is there early."

They entered the store. Frank, the proprietor, came around the counter and shook Rocco's hand, "Congratulations on Mario, everyone is talking about our native son."

"Thank you, I need a couple of boxes of Dutch Masters cigars and a newspaper."

Frank threw up his hands and refused to take his money, "If we had fifty more Mario's, the war would be over in a few weeks."

"You could be right," Patrick said.

As Rocco went out the door, he stopped and lit a cigar. He looked at the front page to be sure that Mario was still there. Patrick noticed Rocco's elation as he took a long drag and blew out a circle of smoke. They walked through the light snow and Patrick

Pico

said, "Enjoy this day, you and you wife raised a fine boy. If my wife and I ever had the opportunity to have a child, I would want him to be just like Mario."

"Thank you, Patrick, I know that you sincerely mean what you're saying. The kid has a way of growing on people."

Before reaching the club, they were stopped by a few pedestrians who wanted to talk about Mario. Rocco shook his head in disbelief and explained to Patrick, "He wrote that the weather was fine and he won some award. Can you believe that guy?"

Rocco banged heavily on the door of the club. The one-legged bartender could be heard cursing inside. Rocco explained, "Tony lost the leg in World War I on the Austrian front. The guy had to be nuts. He was here in America and went back to Italy to fight for the Motherland. He lost a leg and got so small a pension that he refused. He should have fought for the United States. Under the same conditions, he wouldn't have any financial problems."

"That is what everybody says."

Tony let them enter. He had read the paper and congratulated Rocco. He understood this news because he left most of his leg at the Brenner Pass. Tony now produced a bottle of the most expensive cognac. The members owned the club and this was a special occasion. Rocco was an officer of the Sons of Italy Club. Tony had a few drinks and returned behind the bar to resume his work preparation. He made a few telephone calls to some of Rocco's friends informing them that he was at the club.

Patrick and Rocco sat at a far table behind the pool tables. The commander said, "Attorney General Brown called me early this morning about Mario. That's how I found out."

"What did he say?"

"The man was as proud as anyone could be. He feels like a person who is directing Mario from his office and is trying to take all the credit. He told me, 'I told you so.' You know, he genuinely thinks the world of Mario."

"I know he gave Mario the opportunity to clear himself and I greatly appreciate it but anyone, in the present, would be happy to have been connected with Mario."

"I agree with you, Rocco, but do you know what he asked your son when he was leaving?"

Home Town Elation

"No, I haven't the slightest clue."

He pulled Mario on the side so his secretary wouldn't hear him and asked, "Did you really beat the hell out of Johnny Catalano on the bridge?"

Rocco laughed, "Well, if that don't beat it all."

A few friends wandered in and congratulated Rocco. They usually spoke in Italian. Patrick wasn't offended, he loved and enjoyed their company and was accepted as one of them.

Rocco hollered out to the bartender, "If my wife calls, tell her I'm not here."

A few others made the same request. Patrick joined in, "Hey, Tony, if my wife calls, I'm not here."

They all laughed because his wife didn't realize that Patrick patronized the club. Hell, after what Patrick did for Mario, they made him an honorary Italian. Tony had previously called Mr. Hogan. Rocco's friends were waiting for the opportunity to have them mend their differences. They realized that this day was the opportune time. Rocco was in heaven and it was time to forgive Mario's granduncle.

The club hadn't actually opened yet there were over twenty people present when one of the customers let Emilio and Hogan enter.

The small crowd became quiet as Hogan approached Rocco and embraced him. He was actually crying as he said between sobbing, "I'm so proud of Mario."

Everybody was happy to see them make up. Benny said, "Carmela will be relieved, part of her family is back together again." He raised a glass to toast Mario, "Here is hoping that Mario Calcagni comes back home safely."

Patrick Morgan added drunkardly, "I said it before and I will say it again, by God, I feel bad for the enemy that has the misfortune to face Mario."

Emilio found it to be an excellent opportunity to take betting action on the numbers. The bettors sure had enough material to associate with Mario. Hogan raised a glass and complained, "They should have given him the Medal of Honor."

Rocco replied confidently, "Next time."

Hogan announced, "Anyone for a friendly game of pool?"

13

Short Rest For The Battle-Worn

Mario's unit was relocated to a staging area. They were preparing to relieve a similar division in the western sector of Monte Cassino. They spent considerable time training as mountain fighters. The soldiers received special equipment that was more adaptable for rocky landscape.

Mario applied the surveying experience taught to him by Patrick Morgan. He was cognizant of the exact trails they were to travel and studied the maps extensively. Mario understood every square inch of the grid coordinates. He memorized each contour and had an excellent command of the land configuration and rationalized that survival was dependent on knowing the lay of the land. He studied the maps all day long and questioned mule skinners that hauled supplies into the mountains.

Mario had another advantage. His command of the language was beneficial because the haulers were mostly Italian soldiers. Between the maps and vital information acquired, Mario knew the mountains better than most.

He tried to convey his knowledge to his squad members but they had difficulty understanding the contours on the aerial view of a map. He felt, however, that they knew enough to get by. Some of his men complained that he was putting too much emphasis on studying maps. One of his men said, "I bet the lieutenant doesn't know as much about this map as you know."

Mario replied as he folded the map, "If the time comes and he needs my advice, I'll give it to him."

Short Rest For The Battle-Worn

The weather appeared to be the same daily. Rain, cold, wind and rain again. Mario was well-adapted with the existing, persistent weather. New England was comparable and Mario sometimes hunted in bad weather. In reality, he was looking forward to the coming battle. Mario was in his elements. Many soldiers brought down the mountain needing treatment were not battle related. Most of them had trench feet. Mario was determined to get his squad safely through the upcoming battle.

As usual, there was a holdup in logistics. Moving an entire division in mountainous areas is not an easy task. Moreover, they must coordinate with the withdrawal of the existing line division. Mario's group had time to observe incoming ordinance that landed approximately five hundred yards up the mountain. He gave them credit. The Germans had a high degree of accuracy, noticing that the heaviest of the barrages occurred during specific times of the day. Mario calculated, "We've got to move when their observers are eating lunch."

The sad part about it all was that they were concentrating fire on the returning mule skinners. The casualty rate among the haulers was high. One afternoon, Mario and his squad were sitting around discussing tactics when they observed a skinner and his mule some four hundred yards up the mountain. The mule was terrified and was galloping down the trail towards them. His master was hanging on to the animal's tail with one hand and trying to keep a wobbly German helmet on his head with the other hand. It was a comical situation and the Americans began laughing.

When they were near, the soldiers could hear the Italian scream, "Stop Brutus, stop!" There were times when his feet actually left the trail but he wouldn't release the hold from the mule's tail. The animal paid no heed and headed straight towards Mario and his men. Other soldiers in the area were encouraging the mule to move faster. Mario moved into a favorable position and stopped the animal by getting a grip on its halter.

The Italian fell conveniently to the ground and sprawled on his backside. He began cursing which included the name of the Madonna along with every known saint. He wanted some consolation but Mario ignored him and gave his attention to the terrified mule. He patted him on the neck and spoke softly and assuredly, "Nice, Brutus, good boy."

Pico

The animal calmed down as Mario began inspecting its condition. He noticed numerous wounds. They didn't appear to be deep and were caused by flying rocks. The Italian jumped up and complained, "Americano, I'm the one that's hurt. You know the name of my mule but you don't know my name."

"You don't appear hurt to me."

"My pride is hurt. That mule has been making a fool out of me."

Mario played along, "Okay, what is your name?"

The skinner pressed his hands along his wet uniform, gained his composure and stood at attention. "My name is Gaetano Falaguerra. I am from Ciprano and my family is still there, I hope."

"Why do you abuse this poor animal?"

Gaetano threw his hand up in the air and his eyes lit up in approval, "You are Italian."

"Yes, I'm a full-blooded Italo-American and I have relatives who live not too far from Ciprano."

"Where?"

"According to the map, it's called Pico."

"We are paisanos, I've been there many times," Gaetano said happily.

"You know, Sergeant, sometimes Brutus is good and sometimes he doesn't listen to me."

"What the hell do you expect from a mule?"

"Only a little consideration," Gaetano replied.

Mario decided to have a little fun with the Italian as he turned and gave a sly wink to his men. "This mule is wounded, bring him over to that medic station and have him attended to."

Gaetano replied confused, "You are trying to make a fool out of me, I assumed we were paisanos." Mario's men were listening and anticipating what was coming.

"We are friends but I heard that mules are more important here than people. The medics treat the mules at the head of the line."

Gaetano was reluctant, "I know that you're playing games with me, there are no mules in that line."

"That is because it's a small line and you have the only mule that requires medical attention."

Gaetano was weary and asked pleadingly, "Are you sure?"

Mario pressed on. He took a piece of paper from his pocket and signed it, Joseph Paul DiMaggio, Jr. He gave Gaetano the paper

Short Rest For The Battle-Worn

and instructed him, "Just show this to the medic there, you and Brutus will be all set." Gaetano saluted Mario, took hold of his mule and escorted it to the medic station.

The squad was watching as the Italian got into line with Brutus. He was not about to attempt the front of the line. Some of the G.I.'s in the line played along. Gaetano held an extended conversation with Brutus which brought on a burst of laughter. He looked at Mario who signaled him to move along with the line.

When he and his mule reached the front of the line, the medic looked up in complete surprise, "What the fuck is this all about?" Gaetano didn't know and all he could think of was to hand over the note. The men in line were all laughing as the mule dropped a load of manure. Gaetano pointed in desperation towards Mario and then pointed at the note.

The medic looked at the note and then at Mario in the distance. He couldn't forget the sergeant and decided to play along. Mario had saved his life at the river. The medic burst out laughing. He gave Gaetano aspirins and some salve. He instructed Gaetano, "Give the mule two pills, three times a day with water. Place salve over the wounds as often as possible and be sure to give him rest."

The medic was a Brooklyn Dodger fan. He told the Italian, "Go back to Di Maggio there and tell him, 'Fuck the Yankees.' Repeat after me now, 'Fuck the Yankees'." Gaetano got it right, he repeated the statement and Mario had a big laugh.

Mario took a liking to the mule skinner who began hanging around the camp. The following night, Mario and Gaetano were sitting by themselves. Mario was asking the Italian about the terrain. There was very little that he could add but confirmed the information that Mario already had. He asked Gaetano, "Is it possible for a person to get through to Pico from here?"

"Yes, it's possible, but very dangerous. Not only from the war zone but the Germans are taking everybody in sight."

"Why?"

"They are putting many of the Italian men into slave labor. Most are used to build fortified emplacements and others are disappearing northward towards Germany. The other danger is that, the only way to get through is to dress as a peasant. If you are caught by either side, they can shoot you as a spy. It is none of my business, but why do you ask?"

Pico

Mario gave him a cigarette and held a lighted match out to him. "I have close relatives in Pico and wondered if we don't break through, I'll go alone."

Gaetano was surprised, "You are in the American army. Only a deserter would think of such a thing."

Mario laughed, "Thank you for your concern. Sooner or later, if we don't advance, our unit will come back down from the mountain and rest in a rear area. Instead of going to Naples for leave, I may decide to take a few days off and go to Pico."

Gaetano looked at the dark sky, "We are only about fifteen air miles from where you want to go. In peace time it would take about twenty minutes by car. If you walked from here it would take you about six hours. However, under our present conditions, it would take four or five days."

"It was only a thought that has no chance of success. I could probably find a legal way to leave for a few days but not a few weeks."

"I don't think you can leave legally. Your government has spent a lot of money to train you, I don't see how you could ever pull it off."

"I received a battle award and they may allow me some slack."

"I have a friend who did get through. He visited his family near there and came back."

"Are you kidding me? Why in the hell did he come back?"

"Italians are crazy, the asino has a girlfriend in Naples and he won't leave her. Besides, that's life!"

Mario laughed and agreed, "I guess that you're right. It's the same in the States. The Italian men all have girls on the side. I take that back, only the egotistical conform to that statement."

Someone in a nearby tent barked, "Why don't you guys shut-up. We can't get any sleep in here."

Mario replied, "Fuck you and the incoming artillery."

Gaetano ignored the last exchange between the G.I.'s. "You know Mario, I love my wife and family very much and I wouldn't dare keep a girl on the side."

"Why not? Is it because you're too poor or too ugly?"

"What do you mean ugly? I consider myself handsome and dashing especially when I dress for a festival. The truth is I'm afraid of my wife. She comes from a vengeful family."

Short Rest For The Battle-Worn

"What the hell are you talking about?"

"I found out years ago that you don't cross the women in that family I married into." Gaetano became uneasy but he further explained, "My sister-in-law stabbed her husband because she caught him with another woman. That is only for openers. Her aunt cut her husband near the balls as a warning."

"Did she cut them off entirely?"

"No, he was fortunate. His compare got him to a doctor and they saved the family jewels. I don't trust my wife, she may be the worst of all of them."

"If I were you, I would get the hell away from her. Who knows, some girl may make up a story and your wife will nail you while you're sleeping."

"God forbid."

"I know you don't have the ability but do you suppose that your friend can smuggle some putanas up here? It would be real important for my men before we go into the battle areas."

"I'm sure that he can. There are a lot of business girls around and they are beautiful."

"I'll tell you what, if you get some whores here and you fuck one of them, I'll pay the bill."

Gaetano deliberated for a slight moment and agreed. Mario playfully punched him on the shoulder and said, "You son of a bitch, I thought you were afraid of your wife?"

"Hey, paisano, my wife isn't here. When do we get the girls?"

"Never, I was only screwing around."

"I don't know when you're fooling around or being truthful."

Mario changed the subject, "How do you know so much about the enemy fortifications?"

"When Italy quit the war and changed sides, the Germans took Italian soldiers as prisoners and had us building fortifications. The Fascist soldiers stayed with the Germans, but they are no good."

Mario asked, "If you were a prisoner, what are you doing here?"

Gaetano said, "One day when they were marching us up a trail to an area, they all turned left and I turned right and consequently escaped."

"Did you work on this particular mountain area?"

"No, it was the Gustav line, but it was located on the eastern side of Monte Cassino. I'll tell you, the Allies don't have a chance to break through that sector."

Pico

Mario saw a glimmer of hope fade away. If he had worked this area, Mario would have a field day with the information. "Why don't the Allies have a chance in that other sector?"

"Because most of the tunneled emplacements are interconnecting. It doesn't matter how much artillery is fired at them. Their safety is so secure that a direct hit won't phase them out at all."

Mario asked with deep interest, "How about in this area? Are conditions the same?"

"From what I hear and have noticed, their positions are not as strong here but don't let that fool you. These fucking Teduscis are formidable opponents, especially on defense."

"Thank you, my friend, you have been very helpful. I joked with you often but it would be great to see you again when the war is over."

Gaetano was touched and found the opportunity to make a request. "Mario, according to your men, you're a big shot. Can you get me authorization to carry a weapon so I can fight the Germans?"

"If I had more time, I would certainly try but I think we will be moving up within twenty-four hours and there is a lot of red tape involved."

The Italian was downhearted and Mario was concerned with the welfare of his new friend. He took two grenades and gave them to him. "From what I hear, these are more useful than a rifle in the mountains. If you get caught with them, just say you found them." Mario added, "Do you know how to use these?"

"Yes, I do and thank you very much."

Mario rose and bid him good-night as he entered the tent. The Italian slept outdoors under a canvas cover.

The next morning, bright and early, Gaetano got his mule ready. It was time to face the daily gauntlet provided by German artillery. Mario embraced Gaetano and wished him luck. He offered, as a consolation, "Who knows, we may meet again up the mountain. Take care of yourself and be nice to Brutus." As Gaetano left to load his mule, Mario realized that the Italian's chances of survival were perhaps fifty-fifty. He later watched Gaetano and his mule moving in the distant trail and had a feeling that he probably would never see him again.

14

Mario Establishes A Precedent

The word came down that evening: the troops were to make preparations to move up. Mario assembled his squad, "Well, this is it. I wish all you boys the best of luck. Remember, we are going to create our own luck. The only way we can get into trouble is if we cause it ourselves. I don't want any mistakes, not in this arena."

That night, accompanied by wind and rain, they moved out, one company at a time. They halted often because of the constant artillery barrage.

Mario noticed that the higher they traveled, the rougher the terrain. He led the way as Corporal Kelly brought up the rear of the squad. The sergeant had a great deal of faith in his Corporal because he was the most attentive during training. They progressed at a snail's pace and were over half-way to their destination when the rain turned to sleet.

It wasn't snow yet but the wind, along with the sleet, produced difficulty in movement. Consequently, they were all behind schedule. The company was strung out along the rockiest section of the trail. The conditions were so difficult that some stragglers dropped back. Most of them were wounded and others were shell-shocked. Mario ignored them. There was nothing he could do.

The first sergeant ordered from up above, "Move it up, get the lead out of your asses!" They approached a small clearing that was also level. Mario held his squad back and sensed that the clearing was too good to be true. It was dark but the light produced from explosions revealed inert bodies in the small landing. Luck

Pico

was with them, he was correct holding his men back. The German artillery had the space targeted perfectly. The explosive force resulting from four or five incoming rounds shook the entire section. It was fortunate they hit the ground and had some low points between them and the landing. Sections of shattered rocks dropped on and around them. Mario asked, "Is anyone hurt?"

The replies assured him that they were okay. The field phone rang and the first sergeant asked, "Where in the hell are you guys and where are the people who are behind you?"

Mario ordered, "Give me that phone." Trying to speak over the rounds of explosions, he informed his superior, "There are a lot of dead immediately ahead of us and nobody is coming up from the rear."

"Where are you?"

Mario gave him an accurate grid coordinate and the Sergeant replied, "It's safer once you climb another hundred yards or so. You can dig in back there but I don't recommend it. The enemy found out about our movement and they are exploiting the situation."

"We will try to push through, over and out."

The Corporal told the men, "That's a perfect example of making your own luck. Sgt. Calcagni estimated that the landing was dangerous and he was right."

During the first lull from the action, they rapidly crossed the section and picked up the trail along a curving cliff precipice. The enemy was firing mortar rounds on every section of the main trail.

Mario wondered: How do they get the wounded down along here? There isn't enough room for the average G.I.

He scanned the map in his mind and knew exactly where they were. The last heavy barrage had covered the main trail and his squad was disoriented and was moving along an alternate route. Mario calculated that the original trail was approximately fifty feet ahead and above their present position.

They, however, faced a greater risk. The squad was exposed and the terrain was nearly impossible to traverse. They had one thing going for them: The enemy didn't expert anyone in that particular section.

There were times when Mario tied a rope around him to test the footing. It was a relief not worrying about explosions from

Mario Establishes A Precedent

mortar or rifle fire. It was evident they had not been spotted. For the immediate present, they were safe.

It was a different story up ahead as the main trail was being saturated with incoming mortar. They couldn't stay where they were and couldn't move ahead until the barrage let up. Mario waited until midnight approached and rationalized: Perhaps the Krauts will stop for a cup of coffee or a beer. They waited approximately five minutes until Mario thought that it was safe. He knew that most men of his company were probably dug in or sheltered along the main trail but also figured that they were suffering high casualties. The first sergeant had ordered most of them to seek shelter until further orders.

The squad was several yards from the trail's intersection as Mario was still testing the footing when the heaviest of the artillery barrage broke loose. The fire was concentrated on the main trail but an errant projectile exploded close enough sending Mario careening down the steep cliff.

He fell down about a dozen feet and ordered his men to stay as the barrage continued. Mario found some good footing that led to a protective area between a crevice. It was only a matter of time that his squad would be discovered and subsequently annihilated. He called up to Corporal Kelly, "Get everybody down here where it's safe." Within five minutes, they were in the safety of the protective crevice.

The artillery barrage intensified. The light from the explosions revealed some G.I.'s being blown off the steep cliff that was located immediately ahead and to the left. They cascaded and tumbled below Mario's location.

He still had the rope tied around him and instructed his men, "Get a good hold on this rope, I'm going to check for survivors." The sergeant made his way down a treacherous rocky incline that resembled a wall. He landed on a level section and nearly tripped over a dead body.

The brief light exposed another inert body presumably dead. As best as he could determine, they were both officers. He scanned the area between flashes of light. Mario had noticed three men were hit and he could only account for two of them. Chances were that the other was also dead. It appeared that they were victims of a direct hit. Mario was ready to signal his men to pull him up to

Pico

safety when he heard a groan approximately twenty feet below him. A brief flash of light revealed a movement. The surviving soldier was in a precarious position. The only thing preventing him from falling into an impossible rescue areas was a thick bush. Mario ordered, "Give me more slack, I'm going down further."

From studying the maps, he knew his exact position. The right consisted of a ravine which doubled back down the mountain. The enemy was on the opposite side. To his front and left were shear cliffs. The Americans had control of the western side of the mountain. The thing that puzzled him the most was that there was no activity on the opposite side of the ravine. The lack of small arms fire aroused his suspicion. Mario thought: Could it be possible that they pulled back?

He reached the wounded man and to his amazement, he was a Colonel. Mario splashed some water on his face and revived him. The colonel appeared to have broken a leg. His arm didn't look too good and he was badly bruised from the long fall. Mario held the wounded man's head up and gave him water from his canteen, "Don't worry, sir, everything will be fine. I have a safe position about thirty feet above us."

While the colonel was awakening and gaining some composure, Mario was evaluating his position. To his immediate right, he noticed a sharp vertical crevice that continued upwards and closely paralleled the safe section that his squad occupied.

He told the Colonel, "I'm going to tie this rope under your arms and swing you over to that safer area to the right."

The colonel concurred and asked softly, "What is your name, son?"

"Sgt. Mario Calcagni. Now hold on tight with your good arm, I'm going to move you."

The first phase of the rescue was a success. Mario was preparing to tug on the rope as a signal to pull the Colonel up when he noticed movement on the other side of the ravine. He held back and signaled his men, as well as the colonel to be absolutely quiet. The enemy was in the ravine and some of them were getting close to his position. He now realized why it was so quiet in the area and why the artillery and mortar barrage intensified.

The Germans were counterattacking and the ravine was an excellent route to outflank the Allies on the mountain. Mario now

Mario Establishes A Precedent

had an advantage having studied the maps so well. Knowing each grid coordinate, he intended on utilizing the information. Mario signaled Cpl. Kelly to drop the radio by rope. He whispered to the Colonel, "Sit tight here, I've got a little business to attend to."

The Colonel was becoming more alert and encouraged Mario to play it by ear. The radio was on its way down when Mario was discovered by the enemy. They knew he was there but held back their fire. It was understandable, they wanted to make as much progress down the ravine before revealing their intentions. Mario was aware of their plans and immediately opened fire and dropped grenades down the hill. They responded furiously as Mario wedged the colonel safely into the crevice. They had protection from the enemy below but the Germans on the other side of the ravine returned some close fire. The battle raged as his squad from above was covering him by directing their fire across the ravine.

Mario ducked low behind the lower section of the vertical crevice and immediately called for mortar and artillery. The German counterattack was now in full force. The only obstacle was one quick thinking American Sergeant. Mario gave excellent coordinates and announced, "I have a Colonel here with me, so, cut the shit and give it everything you've got."

The first round came in. Mario was in business. He could now direct the incoming barrage and gave more accurate grid coordinates. He ordered firmly, "Give them hell, they are attacking, big time." Mario was catching fire but he had to move out for better observation. The front of the attacking force was halted. Mario had them in a position where they were sitting ducks and kept directing the fire.

The enemy was retreating precariously close to Mario's proximity. He fired his submachine gun and expended all his grenades. His men dropped him more ammunition. Finally, he decided it was safe and signaled his men to pull the colonel up to their position.

Mario began directing artillery close to his position. In fact, he was hit by some shrapnel. Mario yelled into the phone, "You bastards are too close! Move it out twenty yards to the west."

The German pull back was in turmoil. They stopped firing at Mario but he didn't let up one bit. The artillery barrage saturated the ravine and remained constant as Mario made his way back to the position once occupied by the colonel.

Pico

As his men lifted their sergeant to safety he knew the enemy was near annihilation but decided not to call off the artillery. Hell, he said to himself: Those men need the practice.

Corporal Kelly radioed for the medics who were waiting on the rocky trail. Mario and the Colonel were pulled up by his men who were already relocated there.

The Colonel took command of the group and ordered them all off the mountain. It was daylight when the small entourage reached the bottom of the trail. The word had preceded their exploits and a reserve unit lined a path that ushered them along with applause and handshakes. Mario pulled his scarf out of his pocket and wrapped it around his neck. The colonel was beaming as he held Mario's hand, "Thank you, son, you may have saved half of a division." Mario waved his men along in appreciation as he and the Colonel walked to the aid station. Again, Mario was fortunate. His wounds were superficial but the colonel was headed for a hospital near Naples.

Col. Biggs ordered that Mario be pulled out from all future action. His efforts were being hailed as one of the greatest acts of war ever accomplished by an American soldier. Mario was sent back to a Salerno rest area and received the V.I.P. treatment on all fronts. He was called into the General's trailer and was promptly saluted. Mario returned the salute with a hint of confusion, "Sir, why did you salute me first?"

The General came around from his desk and shook his hand vigorously, "Son, you have been awarded he Congressional Medal of Honor and officers are obliged to initiate a salute in your honor. I want you to know that I am honored to be the first officer to pay you that respect."

Mario didn't want to burst his bubble because the Colonel was the first officer to salute him.

The General placed an arm around Mario and informed him that he would soon be going back to the States. The President insisted that he was going to present the award to Mario. The General introduced Mario to his staff. They opened a bottle of expensive cognac in celebration. He enjoyed the experience but didn't relax until he returned to a rest and recuperation hotel back in Salerno. He laid in his bed and took a few swigs from a complimentary bottle of cognac. The radio was playing stateside

Mario Establishes A Precedent

music but all he had in his mind was the thought of his relatives in Pico. He thought to himself: Everybody is happy, including me, but I really messed up. They are shipping me back to the States. I sense problems in Pico and, by God, I'm going there!

After Mario drank over half of the cognac, he said out loud, "Fuck them all! Nothing is going to stop me, I am going to Pico." The music played on into the night as Mario fell asleep.

15

Rest And Recuperation

The nurse entered the room. "Colonel Biggs, there is a Sergeant Calcagni to see you."

The colonel propped himself up and ordered the nurse, "By all means, show him in. Wait, young lady, don't you know who he is?"

"No, I'm sorry, sir, I don't know."

"That is the Sergeant Mario Calcagni who stopped and just about destroyed the German counterattack and also saved my ass."

She replied nervously, "Oh, my God, how do I act?"

The Colonel laughed, "Just act normal when he pats you on the ass."

The nurse fell in love as she said awkwardly, "The Colonel will see you now."

Mario thanked her with a smile and stepped into the room. Colonel Biggs saluted him and, he returned the salute.

"You are looking good, my colonel."

"Yes Mario, I feel better. The leg and arm injuries were not as bad as anticipated. The bruises are completely healed."

"You took one hell of a fall off that cliff. I always wondered, what you were doing in that area."

"Believe it or not, it was safer up above but we were being relieved by your division. We decided it was safe to proceed down the mountain. The Germans caught us by surprise, we thought the enemy had called it quits for the night. You know Mario, I lost my closest aides."

Rest and Recuperation

"Yes, sir, I found them before I found you."

"Son, again, I want to thank you for my life but even more so and I speak with the utmost sincerity, thank you for all the other lives that you saved." The colonel had tears in his eyes and Mario became uncomfortable.

Mario conveniently changed the subject. "Sir, I have a request to ask."

"Mario, the world is yours. You are probably the only man in this army that can ask whatever he wants."

The nurse entered and brought Col. Biggs his lunch. She leaned over him and directed her lovely buttocks towards the hero, expecting a courteous pat.

She left, displaying a perturbed expression and informed Mario, "I'll bring you some lunch."

Mario thanked her as she left. He scratched the top of his head and asked Col. Biggs, "What's wrong with her?"

"I don't know, Mario. Perhaps you should have patted her on the ass."

"Sir, are you giving me advice?"

"Soldier, why don't you play it by ear?"

The nurse came back with his lunch and leaned over the Colonel to adjust a pillow. She was in the same position as before. Mario looked at Biggs who gave him a sign of encouragement. Mario shrugged and he patted her gently. She jumped up, giggled and happily left the room.

The colonel congratulated him, "You made that young lady's day. She will inform the entire hospital staff that the great Sergeant Mario Calcagni patted her on the ass." Mario laughed after realizing that the Colonel maneuvered him into the episode.

Biggs got serious, "What is your request?"

Mario made himself comfortable in a nearby chair. "There is a little mountain town that is located about twelve air miles from Monte Cassino called Pico."

"Yes, I'm acquainted with the name. In fact, the Germans have a very large supply depot there and the road is of strategic importance. Why do you ask?"

"My family is from there. I have a grandmother, two uncles and countless cousins living there."

The Colonel listened with interest and urged Mario on, "What's on your mind?"

Pico

"I'm concerned for their health and welfare. I made a promise to my parents that I will do my best to help them."

The Colonel assured him with confidence, "I will see to it that you will visit and help your relatives after the area is cleared."

Mario pulled out a map from his jacket pocket. He spread it out and asked, "Can we talk?"

"Of course, son. Whatever you tell me is confidential."

Mario traced his index finger along the all-important Route 6. He stopped at Itri and then traced along a rugged road that headed northeast to Pico. He held his finger on the spot and looked up into the Colonel's eyes, "I'm going there before it's cleared."

"What do you intend to do?"

"It's very simple. I intent to do some of the clearing."

"Mario, what you want to do is impossible. The front lines are over for you. I don't think you realize your present status. You are an American hero." Colonel Biggs waved his hands, "There is no way that the General will allow you to place yourself in harm's way." He continued, "The President himself has insisted that he award you the Congressional Medal Of Honor."

"I'm impressed and appreciate everything that is being done in my behalf but I sense that something is wrong with regards to my family." The colonel let out a sigh of hopelessness and studied the map.

"Mario, I'm going to explain some strategy to you. But first, close that door."

The sergeant rose and shut the door placing a No Disturb sign on the knob. Biggs sat up and pulled out a complete map from his nearby folder.

"As you probably already know, there has been a great deal of movement all along our lines. I have been briefed because I've been transferred to headquarters. Two Polish divisions have been moved to the Monte Cassino area. The British have relocated three divisions on the west flank of the Poles. Next on their flank are four divisions from the French Expeditionary Corps. Next on the line are two American divisions. That is the lineup that will be used for the coming spring offense."

Mario studied the map and traced a finger as before through Itri and up to nearby Pico. He said with delight, "It looks to me that our American Division will take Pico."

Rest and Recuperation

"No, Mario, I don't think so. Our main objective is to move as quickly as possible and join up with our forces in Anzio. We probably, in all likelihood, skirt by Itri."

"Well," asked Mario, "who goes to Pico?"

"It appears that it will be the French. To be more specific, the Moroccans have the best chance. They are the best mountain fighters and that entire area from Esperia to Pico is all mountains."

Mario looked up and asked, "Are those the Moroccans known as Goumiers?"

"Yes, I believe you're right. Goumiers or Goums or both."

Mario sat back in his chair with growing concern, "Now I know why I have sensed that something terrible is going to happen."

"Why, Mario, explain it to me. What is wrong?"

"I have heard that these Goumiers are excellent mountain fighters but I also hear that they are animals. I heard rumors that they rape women wherever they are encamped. In fact, I'm surprised you haven't heard. The general himself authorized the importation of Moroccan women to help satisfy their sexual desires."

"You know, Mario, now that you mention it, I do remember hearing the same stories. I know that they did indeed transport Moroccan women here. I don't want to alarm you further but they have been known to sexually attack men."

"When they run out of humans, they take their lust out on animals," Mario added.

"They are valuable for the breakout and I guess that is the reason they are kept strictly in the mountains. The Goumiers are uncivilized."

"I agree. The French are keeping them under wraps but what is going to happen when they overrun Esperia and Pico?"

"I don't know, Mario, but I do share your concern."

"Colonel, all I ask is that you get me transferred to the division that is flanking the Moroccans and I'll do the rest."

"Son, what you ask is not possible. The general will not allow it."

Mario didn't intend a smile but utilized it, "Sir, with all due respect, fuck the general."

"Come on, Mario, that kind of talk is out of order."

Mario was enraged enough to tell the Colonel to go fuck himself but held back. He threatened, "I'm going to that town, come hell or high water. The only way to stop me is to toss this fucking Medal of Honor winner into jail."

The Colonel was searching for reasons of discouragement and replied, "You have about a thirty percent chance of making it there alive."

"You are telling me that the Goumiers are great mountain fighters. Well, I'll tell you, I know a little about fighting in the mountains."

The Colonel retorted, "That is the understatement of this entire war and I personally can attest to that." They both laughed and calmed down. Biggs gave a sigh of relief but he didn't know why. "Mario, what makes you think that you will come out of this alive?"

He became very serious with the intention of letting his colonel-friend in on a secret. "I recently had a powerful dream that was different from any dream that I've ever had. A lady appeared, her image was very sharp. She had a bright halo over her head and told me that I was going to live to the age of ninety-three."

"Mario, do you honestly believe in dreams?"

"Only the ones that I want to believe in, the rest don't count."

"When did you have this dream?"

"The very first night that I stepped on Italian soil."

The intense look on Mario's face told the Colonel that he believed the dream was possible. He thought to himself: Mario had no chance for survival at the river bridgehead and he made it. His situation on the mountain was clearly impossible and he survived. Someone up above is certainly looking over him.

"Mario, the only thing I can suggest is that we both present your case to the general. The chances are that he cannot deny your request. I have an idea that may convince the general."

They were both silent as the nurse knocked on the door, "Is everything all right in there?"

Biggs assured her in the affirmative and ordered her to return later.

Mario was awaiting the colonel's advice. Instead, the senior officer made a statement, "You used a battle strategy that is completely alien to me and I believe that the Germans have made the same assessment of you."

"Sir, I only regret that I caused and was responsible for the death of so many. It is not my character to be a killer."

Rest and Recuperation

Biggs stopped him by raising his hands, "Son, you saved yourself, me and nearly an entire Division of the good guys. The killing goes with the territory."

Mario didn't reply, assuming he will have the answer after the age of ninety-three when he meets the Creator.

The colonel explained, "We should be able to transfer you into a reserve unit that will move forward immediately after the breakthrough. The general is not going to risk his boy's life. We've got to keep it for him."

"Reserve units are not utilized unless they are really needed," Mario countered.

"Not in this case because it is imperative that we move with great speed to Anzio. I assure you, the reserve part of the division will be close enough to draw fire. The main thing is to convince the general that you will be in a safe area." Mario displayed a degree of patience and listened. Biggs continued, "You should be in the Itri sector within a few hours of the advance and I now ask you, what are you going to do at that point?"

Mario's eyes brightened, "I'm cutting out and heading northeast!"

The colonel emphasized, "I know nothing whatsoever about what you just proposed. In reality, you will be A.W.O.L. but who in the hell in this man's army is going to fuck around with Sergeant Mario Calcagni?"

"It will only take me a few days to check things out, I'll be back with my unit before they even know it."

The colonel had a close friend who was a Lieutenant in the reserve unit he had in mind. Not that it mattered but he would advise him to look the other way. He told Mario, "I was going to ask you but now I know why you refused the battlefield commission that was offered you."

"Yes, Colonel, I have a better chance as a sergeant to check on Pico."

"If I wasn't a colonel, I would be honored to accompany you but rest assured, as soon as you get back, you're going home."

"I'm happy that you have faith that I'll be back," Mario beamed.

"You had that dream, didn't you?"

The colonel initiated a salute. Mario embraced his friend and left. He was on cloud nine.

Pico

He returned back to his billets and wrote a letter to his parents. It was tersely worded like a telegram. "The weather is cold and rainy but the country is beautiful. Will soon be heading to your village. By the way, I was awarded another medal."

16

Front Page, U.S.A.

This time, the headlines were different. Mario graced the entire front page.

The reporters had visited Rocco and Carmela the previous night. They explained the information that came over the wire service which was exaggerated but customary. America was starving for heroes and it produced one that was larger than life.

Mario was credited with saving an entire division from entrapment. It was a feat that was rare in the annals of warfare. The news made the front pages of most newspapers in the country. The community was in a state of constant celebration but Mario wasn't present.

Rocco displayed the letter to the reporters and said repeatedly, "By the way, I received another award." Rocco proudly shook his head, "I don't know what I'm going to do with that boy." He and his wife became tired of answering the telephone and decided to leave it off the hook. The Sons of Italy organization decided to stage a celebration and just talk about their Mario. The Easter holiday was approaching and many of the revelers broke their vows for the Lenten season. The festivities were planned to include members and friends only. Mario was still in Italy and a full-scale celebration was being proposed as soon as he returned. Needless to say, the entire program got out of hand.

Patrick Morgan notified the Attorney General and, of course, he showed up. He had a right to be there. However, that opened the door to all the politicians. The mayor arrived along with his aides. The governor came with a police escort. The immediate area was immersed in a traffic jam. Sal Hogan wore a toupee and accepted congratulations from everybody in attendance. Rocco and Carmela required protection from friends who swarmed all over them. Emilio was having a field day. Everyone was betting the numbers.

Benny told Patrick, "I wonder what is going to happen when Mario comes home."

Patrick replied with a big smile, "They better send at least a company of soldiers to protect him."

"Yeah," joked Benny, "they better back it up with Marines."

The band played. American flags were displayed throughout the area. Children were waving miniature flags. Some people thought that Mario had arrived and wanted to see him.

A master of ceremonies yielded to Patrick Morgan, who spoke warmly from the heart. He introduced the politicians who made brief statements. The governor thanked God for Mario Calcagni.

Wine flowed like water. The festivities continued into the night. It was April and fireworks were displayed in the dark sky. It was midnight when Rocco and Carmela finally made it home. He appeared to be worried as Carmela asked, "What is wrong with you?"

"My dear wife, don't you see what is wrong? These people are going to drive our son crazy."

"I know what you're saying but everything wears off. Mario can and will adjust."

"Carmela, there could be another little problem."

"What problem, Madonna mia, do you know something that I don't?"

"We both have the same information. Mario said that he is going to Pico."

"So? That shouldn't take too long. He can pay his respects to all our relatives and then come home."

Rocco was getting a little impatient, "Pico is occupied by Germans."

"They won't be there when he arrives."

"Who gave you that stupid information?"

"Rocco, if the Germans know he is coming, they will pull out."

"Carmela, you and that Irishman are both crazy. The best thing is to keep saying the rosary."

Front Page, U.S.A.

They couldn't sleep as they lay in bed. Rocco said, "Can you really imagine what Mario accomplished?"

"Yes, he saved a lot of soldiers."

"That is not all. He killed a lot of Germans and saved the life of a colonel."

"Did he do all these things alone?"

"No, he had a little help. The reporters said that he had a squad of men and insisted that they each receive the Silver Star Award."

"You mean that he and these few soldiers of his killed over five hundred Germans?"

"No, Carmela. He had some help from the artillery people."

"I wonder what is Mario going to do when he settles down after all the commotion?"

"I don't know. He is a very important person and should be able to do anything he wants. Carmela, go to sleep, tomorrow is another day."

"I can't sleep. I hope our son is safe."

"Don't worry about him. I'm confident that he can take care of himself."

17

French Expeditionary Corps And The White Paper Act

A blue, red and white banner flowed briskly in the strong mountain wind. The French General faced the fresh breeze and inhaled three deep breaths. He turned and faced his Moroccan Goumiers. The moment of vindication had arrived. There was no question in his mind, with a little incentive, victory would be assured. The General, small in stature, became an imposing figure when he began to speak.

"Beyond these mountains that we occupy are the location of a formidable enemy. They are extremely qualified and heavily fortified but you are superior soldiers and better adapted for mountain combat." He slapped the side of his leg with his riding crop for emphasis and continued, "You will move rapidly and kill the enemy. Once you have penetrated the line of resistance the populace is yours. The area beyond is abundant with women, wine and booty."

The Goumiers yelled in approval as he waved his crop in the air. "Your General proclaims that, those women, all the wine and the loot are for your pleasure for a period of fifty hours." He paused as the combatants again yelled in complete approval.

After a few moments, the General continued, "If we are victorious, we deserve it all. On this day, your General swears and will keep his promise to you if you obey and complete your obligations." The Goumiers drew their long knives from their sheafs and hailed both Allah and their beloved General.

Thus was initiated the infamous edict that the Italians called "The White Paper Act."

French Expeditionary Corps and The White Paper Act

Captain Bellanger turned to his young lieutenant with a concerned expression, "How in the hell are we going to control these bastards after that proclamation?"

"Sir, with all due respect, there is no way."

The captain and Lieutenant Bovin were close friends for years and were well aware of the oncoming events. The mountain warriors continued their elated chant as they thrust their long knives skyward.

Capt. Bellanger continued over the din, "Our great general didn't need this edict as a license for rape, murder and atrocities. His battle plan is ingenuous for victory. The incentive wasn't necessary."

"I agree with you. The civilians, for all intent and purpose, are our allies. How can we subject them to the despicable fate that awaits them?"

The Captain kicked a loose rock in disgust, "I wonder if the Fifth Army commander is aware of this edict?"

Lt. Bovin shrugged, "What difference does it make? The commander favors our general and, in war, everyone wants to win."

The Captain raised his eyebrows, "To win at any cost? To destroy an unsuspecting populace that has hopes of liberation?"

"We have air dropped pamphlets informing these people that we are coming and asking them to assist us in defeating the enemy."

"I feel sorry for them and hope that God forgives us."

The General departed as the Goumiers quieted down. They sustained an emotion of silent frenzy. Lt. Bovin made an observation, "Captain, I see the overall picture. We can break the resistance line without the Fifty-Hour Edict but the General wants every card stacked on his side of the deck."

The colonel approached as they saluted him. He returned the salute and ordered, "Get your company ready as planned but we will be delayed a few hours because supplies are lagging."

"Sir, how do we handle the fifty-hour period?" Capt. Bellanger asked.

"Captain, you are required to follow the orders from the general himself. Do you have a problem?"

Reluctantly, Bellanger replied, "I never question the general but what happens after the fifty hours have elapsed and our Goumiers are still out of control?"

Pico

The colonel was not perturbed, "The time period is only a formality, we can put a stop to it whenever we deem fit. Five days, ten days, who gives a fuck? If the situation gets out of hand, shoot the bastards."

The senior officer left as Capt. Bellanger shrugged and emphatically recommended to his Lieutenant, "Be sure you have enough ammunition."

The French Expeditionary Corps was well prepared for the upcoming battle. They were leagues ahead of their allied counterparts in mountain warfare. Instead of motorized mobilization they cleverly optioned for use of pack animals. The Moroccan contingency moved supplies into areas that were considered inaccessible. To press and fortify their jump-off position even further, they carried their own supplies which gave them a distinct advantage. General Alphonse Juin devised a plan that could not fail. His point of attack with his Moroccans was concentrated in an area offering the most difficult terrain. The Germans never considered an attack from such an impossible sector and consequently gave very little consideration for defensive purposes.

The general had his mountain warriors concentrated in an area of minimum resistance. History has revealed that the Goumiers had a relatively easy time fighting an enemy that was nearly nonexistent. Their greatest obstacle was the mountainous terrain to which they were highly adaptable. The question that remains is: Who was the enemy, the Germans or the populace that were going to be subjected to modern day atrocities?

The Italians of the various villages immediately north of Monte Cassino, can reluctantly answer the question. One of the greatest contributions generated by Greece to western civilization is the connotation of the word "Why."

From the days of Biblical times to modern-day history of war, the atrocities of the Holocaust is the only horrific episode that exceeds the wrath inflicted by the Moroccan Goumiers on a friendly populace.

18

Devils Released From Hell

The main phase of the battle of Pico finally arrived. Turmoil, understandably, existed in all areas. The civilians were terrified as the front lines changed on an hourly basis. They understood their perilous position but were looking forward to liberation from the Allies. Whenever the advancing soldiers occupied the surrounding sectors, the Germans counterattacked and re-occupied previously lost positions. After several days of intense fighting, the Germans yielded and the allies were finally successful.

The civilians were waiting in great anticipation for the Americans. They encountered an experience in the ensuing days which dwarfed the suffering but the Americans were nowhere to be seen.

Instead, they met a horde of liberators who displayed a much greater taste in committing atrocities than fighting their enemy. They were dark-skinned and dirty. Their uniforms were robes of brown with vertical stripes. Their method of communication resembled the sound of squealing pigs. They flashed long sharp knives which conveyed and established a fear into anyone they came into contact with. Needless to say, the civilians were caught by ultimate surprise. They heard rumors that "bad soldiers" were coming but could never prepare for the events that were to follow.

Women were screaming throughout the night along with the squealing of the attacking Moroccan Goumiers. The cries were terrifying. There was no let up, the raping and acts of sodomy continued with no respite.

Pico

The Germans had counterattacked through the farm that was occupied by Anna, her family and the group that took refuge there. The Germans chased them all out and they relocated to an adjacent property when the Goumiers arrived. They were looking for the youngest girls first. The civilians were alerted to their attentions. The old women of the group ordered the youngsters to run and hide.

Anna called out terrified to Gabriella and Angelina, "Run, hide in the stalls of the pig pens!" There were no animals left but it could be a good place to seek shelter. As the Goumiers were breaking the front door down, most of the young girls had escaped out the window. Anna confronted them and pleaded, "Take whatever you want, please leave us alone."

Before she could continue, she was hit in the face and thrown to the floor. The Goumiers who struck her jumped on her and placed his knife at Anna's throat. Remaining civilians retreated cowardly to the rear of the room as Anna was raped. She didn't scream or cry. It would be to no avail. The other Goumiers selected other women in the room and beat them to submission.

They let out their familiar squeal and raped the women in the presence of the old men and children. Anna cried out for them to leave but was struck again and raped repeatedly by other Goumiers who entered the farmhouse. They were not satisfied and realized that all the young girls were missing and began searching the barn and stall. Angelina and Gabriella were hiding in the darkness of the last stall. They could hear what was happening. The Goumiers were discovering the girls and raping them at will. The ones who resisted were hit with the butt of a rifle and raped repeatedly.

They left a trail of blood. Some of the girls died from the beatings. Gabriella was terrified as Angelina whispered to her, "Remember how we used to run in the field? I would win every time." Gabriella nodded between sobs. Angelina continued, "We must get out of here and I want you to run faster than me. Gabriella, tonight, you must win. Our lives depend on it."

They could hear the cries and squealing and knew that their time was coming. Angelina was worried about Gabriella and shook her, "Get a hold of yourself. We are going to make a run for the hills near our house. We can hide and they will never find us because we know where and how to hide."

Devils Released From Hell

Gabriella finally retained some of her composure, "I'll be all right, I'll follow you."

"Someone is sure to see us. We need to get a good head start. Don't be too afraid because we are traveling light. They are carrying weapons and heavy clothing. The other thing is, we know where we are going and they have no idea." Angelina continued, "Who knows, we may get lucky and they won't chase us?"

Gabriella pleaded, "God, please help us."

Angelina comforted the older girl, "God is very busy right now. We must do our best to help ourselves."

"Angelina, I'm so afraid."

"Don't worry." They embraced tightly. "When I say move, we run, but first, we must crawl quietly out of here."

Gabriella agreed and she followed as they crawled in the mud. After reaching the outer limits of the pen, they were seen by the Goumiers.

"Run, Gabriella, run! " Angelina yelled.

The Goumiers gave out terrifying squeals. Four of them chased the girls as they sped to the trail. Angelina was holding her cousin's hand and was pulling her along. Gabriella was running as fast as she could but was holding Angelina back. They had a sizeable lead on them and were able to slip unnoticed down the side of a hill. The Goumiers approached and, having lost sight of them, concentrated their search along both sides of the trail. Angelina and Gabriella cuddled together but their chasers wouldn't leave the immediate area. They were determined to find the girls.

Gabriella whispered with a soft but scared voice, "I wish that Horst were here."

Angelina covered her mouth with her hand and was successful in muffling her voice. She whispered in her ear, "We can't stay here. We've got to run for our barn."

Angelina said, "On the count of three, follow me and run like you never ran before."

Gabriella nodded in assent.

"Okay, one, two, three!" Angelina sprang and began running away from the edge of the trail. Gabriella followed on cue but tripped over a rock and cried out, "Angelina!" The younger girl turned in fear, looked back without breaking stride and noticed that a Goumier had her in his filthy grip. Another stayed behind as two pursued Angelina.

Pico

Gabriella kept screaming for what seemed like a few minutes. Suddenly, the cries subsided. They had beaten her into submission. Angelina held back the tears, she was busy trying to escape. She cut through the small ravines and crevices in the dark.

The mountain fighters were having difficulty keeping her in sight. Angelina had never lost confidence, she could find her way around the mountains with her eyes closed. The Goumiers could only follow with the sound of moving brush. She thought to herself: If Gabriella didn't trip, she would have made it. Angelina took a curving route and came into view behind the barn.

The Goumiers were so far behind her that it became an advantage. They were on the top of the hill when they saw her in the bright moonlight entering the barn. They were delighted. How convenient of the little girl to lead them to the comfort of a barn. Angelina climbed up to the hay loft and moved quietly while searching for the hidden shotgun. She found the weapon and let out a sigh of relief.

"Thank God for that good Tedusci." He took a chance in giving the gun back to the Micheletti family and had taken an even greater risk in teaching her how to use it. Angelina had played ignorant; her grandfather, Antonio, had instructed her how to operate the weapon. A tear came to her eyes, "God rest his soul."

Angelina thought: If they don't find me, I'm lucky. If they do find me, then they will be unlucky. It was cut and dry. Kill or be killed.

The two Goumiers took their time getting to the barn. They thought they had her trapped. She cracked open the double barrel shotgun and removed the shells to inspect the barrels. Pointing the weapon to a clear open window revealed them to be clear of any obstruction. Angelina replaced the shells back into the weapon and closed it. She made herself comfortable with her back against the far wall. The ladder that came up to the hayloft was about twenty feet away from her. Angelina was well concealed in the darkest area of the loft. She placed the weapon across her lap and waited. As time passed, she was convinced that the Goumiers had lost her. She said to herself: Poor Gabriella, I hope they didn't go too far with her. She hasn't fully recovered from that voodoo fattura. Now this has to happen. How about my brave mother? She will surely get herself into trouble. These are not people, they are animals and talk with them is useless.

Devils Released From Hell

In a short period of time, many thoughts raced through her mind: If only we had gone to America. My mother wouldn't leave her parents and now, they are both dead. If only we had taken up Horst's offer to hide in the high mountains, we could have avoided this situation. We should have gone with the Calcagni brothers.

The little girl shrugged to herself: It doesn't matter anymore, only the present and the future are significant. Angelina pushed the safety forward into the firing position. The safety was off and the baby cannon was ready.

She tried not to imagine the atrocities that were occurring. Shooting was still dominant in the night. An occasional cry would sometimes pierce the stillness. Angelina heard the barn door creek. Someone was entering but it could be anyone. She thought: Possibly Gabriella or Anna?

Angelina would normally call out but her instincts prevailed. She kept absolutely quiet. The girl was cold and calculating. The brave Angelina witnessed many hardships in the past few years and was a formidable foe for any enemy. The Goumiers gave each other a confident squeal. Angelina heard their primitive form of communication and realized what was about to happen.

They searched around downstairs as she could hear the swishing noise of their long knives. They were playing a game because it was obvious they knew she was in the hay loft. The Goumiers looked up and then at each other. They began climbing the ladder together while occasionally stopping to squeal. Angelina waited patiently. They were careless and had no idea what awaited them.

She had a plan. One must be allowed to get close so that the other was also up on the loft. Angelina had two shots; one for each of them. There was no way she could miss at such a close range and consequently disregarded the expected force of recoil on her small shoulder. The shooting picked up in the surrounding hills. It was possible that the Germans were counterattacking. The two Goumiers were no longer interested in the their enemy but Angelina assured herself: I'll show them who their enemy is.

The first one reached the top and playfully used a squeal as a form of whistling. He couldn't see her but she could see him and had his form in her sights. The little finger was on the front trigger as he came forward. He looked in all directions and was taking his time.

Pico

Angelina kept him in sight but was patiently waiting for his cohort. The first one was no more than twelve feet away from her when she observed the full outline of the second one. He was on the top rung and awkwardly lifting himself upwards. Angelina made a split-second decision that raced through her mind: Now is the time. The first was in clear view and the second was out of position.

She fired at the first Goumier as the flash from the deafening blast produced a blinding light. He fell forward as Angelina's finger slipped down to the second trigger and fired a clear shot blasting the other Goumier completely off the loft. They were both dead.

Angelina cast the shotgun aside. There was a smell of gunpowder as she removed the long knife from one of the dead Goumiers and pierced his body in the stomach. After making her way down the ladder, she kicked the dead one that was sprawled on the dirt floor. Angelina paid him the same compliment. She pierced the long knife into his stomach.

The time was probably three in the morning as Angelina looked outside the barn. The shooting persisted in the mountains but seemed to be farther away. She scanned the area but didn't notice any activity. Angelina was concerned about Gabriella and feared that the other two Goumiers may be close by. She went back into the barn and retrieved one of the weapons.

After being satisfied that the coast was clear, she darted to the farmhouse. The door was broken and the house was dark and empty. The little heroine climbed up into the attic to her favorite hiding place and fell asleep with the gun by her side.

Unknown to Angelina, her mother was raped four times. During the night, additional Goumiers came to the house. There were not enough women to go around that they took turns with the same victim.

Some of the women, including Anna, escaped and hid in the nearby hills. The grandmother Calcagni who could hardly walk was raped six times. When there were no longer any women left, they turned to the men and sodomized them. The devils had lost all interest with the war at hand, they were having a great deal of pleasure at the expense of the ill-fated civilians who awaited liberation. It was early morning when Father Notte was saying Mass for the nuns who taught school. They thought they were safe in church. To avoid bombing and shooting, it was common practice to seek shelter at museums, churches, etc.

Devils Released From Hell

The Germans had allowed the priest and his staff to remain in Pico. During the middle of the mass, the front door opened and a horde of Goumiers rushed in with their terrifying squeals. Father Notte turned and faced them. They had no regard whatsoever. The first part of the group swarmed all over the nuns and stripped them.

Father Notte could not believe what he was seeing. He yelled over their primitive screeching, "This is the house of God. This place is sacred!" They paid no attention and even fired a few shots into the ceiling. They began raping the nuns as the priest was crying and pleading for help from the Almighty, "Please God, not this way."

The nuns were beaten and raped as Father Notte tried to intervene. One of the Goumiers crushed his head with the butt of a rifle. The priest died immediately from the blow. After an hour of completing their pleasure, they departed and left behind six nuns who were near death and a priest who was lying inert in a pool of blood.

As the uncontrolled band of Goumiers left the church and entered the street, they were confronted by Lieutenant Bovin who had heard of the atrocities and was trying to restore order. They shoved him against a wall and placed a knife at his throat.

He was threatened with death. One of the leaders struck him in the midsection with his rifle butt. The Frenchman fell to the ground groaning. He crawled and made his way into the church. He feared they may return and kill him. When he saw the scene in the church, he fell to the floor and cried in despair while pounding his fists on the floor.

Dawn was breaking as Angelina was awakened by a noise in the house. She looked through a crack in the attic floor and recognized Gabriella distraught and trembling. Angelina saw that she was alone and went down to comfort her. Gabriella screamed. It appeared that anything would frighten her. Angelina said, "It's me, you are safe now."

Gabriella cried so much that her eyes were red. She was pale and weak, she pleaded, "I need water, I've got to wash the filth from my body." Angelina produced a basin of water and helped her cousin cleanse herself.

She gave her some water to drink and convinced her to climb back into the attic. "We will be safe here as long as we are quiet, soon, everything will pass."

Pico

Gabriella replied, "When? It's already two days of hell."

"Rest, you have been through too much and I promise you, it won't happen again."

"Angelina, I'm so fortunate to have you as a cousin and a sister. You are the only one I can talk to."

"That's not true. Horst loves you and you will someday be able to talk to him all day long."

Gabriella bowed her head, "For all I know, Horst is probably dead and even if he isn't, he won't ever want me."

"You're wrong, Horst's love for you is strong. He killed two people because of what they did to you and he proved his love when you seemed all but dead. Horst never gave up. Somehow, he knew that you wouldn't die. I know that brave German well. He will be back for you and his love will be stronger than ever."

Gabriella began getting some color back. She pursued the topic, "Do you really think he will feel the same after he hears what happened to me?"

"If Horst were here now, there would be a lot more dead Goumiers around. Get some sleep, I will be here near you. I have a weapon and I know how to use it."

Gabriella was relaxing as Angelina confided in her. "In one way, you are very fortunate, you have Horst and people like that Teduschi don't grow on trees. I hope that someday a person like Horst will enter my life."

Gabriella squeezed her hand and forced a smile, "You are only fourteen years old, strong and a special person. There is no doubt in my mind that someone even stronger and very, very special will come into your life."

"Do you really think so? The Calcagni brothers are too old for me and besides, one of them is married."

"I am positive that someone will soon be entering your life."

"What makes you say that, Gabriella?"

She smiled confidently for the first time, "I had a dream about you."

Anna was hiding in the hills with a group of women. She said, "I've got to look for my daughters." Her friends tried to discourage her, "Stay here. It's dangerous. Besides, there is nothing you can do about whatever happens."

"If need be, I will give myself up for my daughters to keep them pure."

Devils Released From Hell

"Anna, people will understand what we have been through."

She looked at her friends with despair, "You are sure of that. How about Gracia, the girl who had a baby with that German officer a few years ago? You people made her an outcast." They didn't reply. She continued, "He wanted to marry her and she refused. I wonder if any of you people are going to talk to her now. Everyone of you, including myself have been raped by these filthy animals. Your daughters have had the same experience but I'm not giving in. I'm going to find them and protect them."

The beatings, raping and plundering continued. The Goumiers were supposed to proceed northward, but they defied orders and lingered in the area. They made themselves at home. Doors were removed from farmhouses and used for firewood. Many fruit trees were destroyed and utilized for cooking fires. They foraged for rape victims as normal soldiers foraged for food. It became more and more difficult to find women and children so they did the next best thing. They raped the men. When through, they usually killed them. There were incidents reported that they raped animals.

The American High Command was notified of the atrocities and ordered the French Expeditionary Corp commander to immediately halt the gruesome activities. They tried to stop something they should never have allowed in the first place. There were many cases during the final stages of the atrocities that French soldiers and officers were shooting the Goumiers in the streets of Pico. Some even faced a firing squad.

Anna left the safety of the hills and proceeded to look for her daughters. She reasoned they had to be close to their farmhouse. It didn't take her long to reach the main road that led to her home. Anna was a few hundred yards from her destination when three Goumiers confronted her. She tried to explain that she was raped the previous half-hour.

They either didn't understand or didn't care and proceeded to attack her on the small road. The Goumiers squealed and grunted as they took their pleasure. The last one to rape her had a grenade on his belt that she secretly removed. They must have had some compassion because they didn't beat her. Anna was still prone on the ground as the satisfied enemy walked confidently down the road. She pulled the pin on the grenade and rolled it down the hill while covering her head. The nearest Goumiers heard a noise and

Pico

assumed it was a stone. The grenade found its mark and exploded between them. All three died as Anna looked up and said aloud, "You will never rape again."

The sound in the mountains carried very well. Anna could still hear cries and screams. The shooting subsided which indicated that the Germans had finally given up the area and were retreating to the San Giovanni sector. She thought: If only the Americans will arrive.

Anna reached her house and peeked in. The place was empty and quiet. Gabriella was still asleep as Angelina cocked her weapon and aimed near the open door area. Anna stepped in and Angelina cried out softly, "Mama." Gabriella woke up as Anna cried out with joy, "My two daughters are safe." Angelina came down and put a finger to her lips, unseen by Gabriella. Anna understood.

They were all embracing each other. Gabriella held her head down. Anna placed her hand under her chin and raised her head, "We must all try to forget what we are going through. You girls have a long life ahead of you. Wipe out everything you've seen."

"You don't know what its like to be beaten and raped," Gabriella said.

"Yes, I do, my sweet Gabriella. I've experienced the same things that you have. Look at my bruised face. My entire body is bruised and I carry an odor that I abhor."

Angelina felt sad for her mother and embraced her. She cried, "My poor Mama, what have they done to you?"

"They have done to me what they have done to everybody they have come into contact with."

Anna began washing herself as Angelina looked out the door with the weapon in her hand. Anna said with concern, "If someone sees you with that gun, you will be shot."

"Mama, nobody is going to touch me and nobody is going to violate either you or Gabriella ever again." Angelina continued, "We just heard an explosion nearby. It happened just before you arrived here. I think they are coming back."

"Don't worry about that explosion. There are three less Goumiers to worry about."

"Why, Mama, what happened?"

"I gave them a grenade to remember me by."

Angelina said, "That makes five because I shot two of them in the barn. The Micheletti family is taking care of business."

Devils Released From Hell

"How did you shoot them?" Anna asked.

"With the shotgun that Horst gave us."

Anna made the sign of the cross and said, "Thank God for Horst. I hope that he is safe."

During the rest of the day, some of the original group made their way back to the Micheletti farmhouse. One of the old men replaced the door and they tried to make the best of a bad situation. Horror stories were being related but Anna stopped them, "We must cease talking and do something to ensure our safety."

The group quieted down and asked Anna, "What do you have in mind?"

"They will be back as soon as they get some rest and we are here like sitting ducks. We must move in the mountains. The shooting has nearly stopped. The war with the Germans is over, and the only war that continues is our confrontation with these animals."

One old lady said, "They can't stay here forever. The Americans will come." Anna threw up her hands in disgust, "The Americans are coming, the Americans are coming. Forget it, when they do come, all well and good, but until then we must move around and hide." They all agreed with Anna. She told them, "Tonight, as soon as darkness arrives, we leave."

They packed what few provisions they had hidden and waited for darkness. Anna, for the first time in months, thought of her husband, Angelo, in America. She knew that he realized in later years of his grave mistake. She reminded herself: Her father was dead. Her mother was dead. She was raped. Gabriella was raped. There was nothing untouched except Angelina and she was concerned about her. Horst exposed the girl to vengeance and killing. She now has killed two Goumiers.

Anna covered her face with her hands, "My little jewel, my little Angelina, I pray to God that you are strong enough to recover from this terrible episode of your life."

Angelina came to her mother, "Don't worry, Mama, we will make it."

Anna hugged her like a teddy bear and looked into her eyes, "I am so very proud of you. I want you to grow up and be a fine young lady."

"I will, Mama, I will."

"Thank you, Angelina. Now get rid of that weapon."

19

Mario's Trek To Pico

The Allies were finally on the move. They broke the Gustav line. The French Expeditionary Forces penetrated the difficult mountain terrain that the Germans left sparsely defended. They didn't believe that anyone could move in the treacherous and impassable areas. The Germans also underestimated the fighting ability of the French, most notably their mountain fighters. They cut through with force and caused the collapse of the entire defensive line. The redeployment of German defenses to plug the French Expeditionary Corps' breakthrough allowed the Allies to move on their own accord. After a few days of fierce fighting, the Germans dropped back to the Hitler Line. They kept counter-attacking as they retreated.

Mario's regiment was in reserve but moved along the advancing elements. They reached the Itri road when Mario met up with the lieutenant who was a friend of Colonel Biggs.

Mario shook his hand and said, "Well, here is where we part company."

"I don't advise it, but who am I to tell you what to do?"

Mario was stacking up grenades and packing extra ammunition clips for his submachine gun. "I will catch up with you in a few days. I don't expect any trouble. By the time I get to Pico, the fighting will be over."

"I don't know, Sergeant, the Germans are giving up ground grudgingly. They counterattack whenever the opportunity arises."

The lieutenant reviewed a map with him. He noticed that Mario was going to move through the new Hitler line.

Mario's Trek To Pico

Mario said, "I'm not worried about it, one man can filter through unnoticed. I have studied every square foot of terrain and will put the knowledge to use."

The officer wanted to remind Mario that he knew nothing about his trip but realized that it was unnecessary.

He told the Lieutenant jokingly, "Keep an eye on things, I've got to take a piss."

"Hurry back, because we are moving out."

Mario disappeared down a trail that ran parallel to the road. It was dark except for the occasional flashes from exploding artillery. The rapid fire from automatic weapons reminded Mario that he must be alert. He traveled slowly for approximately three hours. The dawn arrived as Mario had skirted the town of Itri. He estimated that he was on a road that coincided with the position of the Moroccan division, using the road as little as possible as the American kept mostly to the dense trails.

There was a great deal of action all around him. Indeed, the Germans were retaliating. The front lines were unstable, but Mario continued to move in a northeast direction.

After completing a rough section of rocky terrain, he soon heard conversation coming from up ahead. Mario stopped as a precautionary measure. English voices appeared to be pleading, while the French seemed to be ordering. There was a clearing ahead and Mario approached. He moved quietly and undetected while reaching a position that revealed a mixed group of ten soldiers. Six of them were clad in long brown robes that covered some type of military uniform. They wore helmets that did not resemble the American issue.

The other four were being treated as prisoners and were without their helmets and weapons. Mario evaluated the situation. He was positive that the prisoners were American soldiers, but he couldn't understand what was going on.

He advanced without being noticed and was soon able to hear the conversation clearly. Mario understood the English, but had a difficult time trying to determine the other language that was accompanied by grunts and groans. He recognized the language to be French but had no idea what they were ordering. The one thing he could determine was that they were up to no good.

Fifteen minutes passed. The soldiers in robes ushered the Americans to the far edge of the small clearing and noticed that

Pico

they were being told to take their clothes off. The Americans complied as they were jabbed with long knives.

Mario wasn't sure who they were, but now he was positive. They were the Moroccan Goumiers. Everything was quickly falling into place. These guys were planing an act of sodomy. They laughed, grunted and squealed making no sense whatsoever. Mario wanted to be sure as the aggressors made advances towards the unarmed Americans. When positive of their intentions, he sprang out into the open and now had a clear view of the Goumiers. They were all caught by surprise and Mario wasn't looking forward to an explanation of a language he had no knowledge of.

He didn't say a word, but the look in his eye told the story. A few of the Goumiers moved their weapons toward him to fire.

That was all that Mario was looking for; the indication that they were going to shoot him. He fired his submachine and felled all six of them. The unsuspecting Americans were startled. Mario approached the downed Goumiers and noticed some movement. He leveled his weapon and finished them off. He ordered the Americans to get dressed. They were relieved. One of them said, "Thank you, but weren't you being a little too hasty?"

Mario was kicking each body to be sure they were all dead. He looked up very slowly and momentarily hesitated, "What the fuck are you, a queer?"

The soldier was petrified but managed to answer, "No, Sergeant. Why do you ask?"

"You dumb bastards. Where the hell were you guys reared, in a bag?" They didn't answer.

Mario calmed down, "Those dead fuckers are Moroccan Goumiers. They rape everything in their path."

"You mean that they were going to pull something like that on us?"

"They were not only going to fuck you, but they were going to kill you when they were done. Haven't you boys ever heard that dead men tell no tales? You were going to be the dead men. I came along and reversed the role, they're dead."

The soldiers dressed and gathered their weapons. They each shook Mario's hand and thanked him. Mario advised them, "Don't allow what just happened bother you. This is war and they got out of line." Departing, one of the grateful men asked, "Who do we owe our lives to?"

Mario's Trek To Pico

"The name is Mario Calcagni. Now get back to your outfit." Mario began walking in the opposite direction as one of the men yelled out, "Hey Sergeant, you're heading in the wrong direction." Mario waved his hand in the air without turning. He was on the road heading towards his destination but used the northern trail boundaries as much as possible.

The evidence of war appeared to be restricted to the mountains north of the road. The Germans were still in control of this sector. They were making a determined effort to hold fragments of the Hitler line. Mario knew that sooner or later, he was going to run into the enemy. They were retreating and they were doing it in force.

Mario knew where he was at all times. Maps were his hobby and he got great satisfaction identifying a massif or ravine. The closer he got to his objective, the greater the sounds of explosions. He feared the worse. Pico was under heavy attack. He was hoping that the British Eighth Army would somehow deviate from their boundary with the American Fifth Army and take Pico. The sound of the battle clearly confirmed that the French were making progress in the area.

The Germans were utilizing the road to relocate units to plug sparsely defended sectors. It appeared that they were already outflanked because there was shooting far to the northeast. Mario waited for the right opportunity and crossed to the northern side.

He wanted to pass through a small mountain top hamlet known as Campose Mele where his grandfather and his granduncle Hogan were born. The small community was only about four miles from Pico.

Sporadic shooting continued as he made his way up the trail. Mario stopped and studied the map while he had the advantage of daylight. The trail was winding and he halted on numerous occasions. Evening arrived as he reached the edge of town, a small hamlet, consequently, with few people.

The Goumiers wouldn't waste their time with low numbers. The Germans didn't need it, this was of no strategic value. He determined that the place was safe. The Moroccans had passed through the area and there would be no counterattack.

Mario waited and scanned the few buildings that resembled small huts. Some were on fire and there didn't appear to be anyone around to put out the fires. He thought the hamlet is located near

Pico

the very top of the mountain and water must be scarce. The American walked cautiously into the little village and it was deserted. The inhabitants probably ran off and were hiddden in caves.

He spotted an old man searching through the rubble of a destroyed hut. The old man didn't see him and was startled when Mario called out softly, "Paisano."

The old peasant tried to run as best he could as Mario called out to him in Italian. He stopped and trembled as Mario approached with an assuring smile.

"Oh," he discovered with relief, "You must be an American."

"Yes, but don't get your hopes up too high because I'm alone."

The old man became at ease and smiled. He tugged on his old suspenders and snapped them with delight, "My name is Antonio Ponti and I have relatives in America."

Mario smiled. "Nearly everyone in Italy is named Antonio and they all have relatives in the United States."

Antonio invited him to his little farmhouse and explained that the Moroccans missed him when they hit the hamlet.

"You were fortunate, how did that happen?"

"I guess I was on the right side of the hill. My wife has a little altar in the house and prays all the time." Antonio moved along the steep trail effortlessly. Mario noticed: Antonio may be old, but he's in great shape.

They reached the house which was larger than others in the village. It consisted of a thatched roof and the sides were made of straw fortified with mud. He thought to himself while shaking his head in disbelief: How could one of the best pool shooters in the East Coast be born here? He almost started to laugh: That crazy Hogan, I don't believe the population here is over fifty people. Of course, that's not counting the goats."

The house had a small makeshift barn and a suspect corral. Antonio stopped and proudly announced, "This is my home."

Mario congratulated him with sincerity, "It looks comfortable and it's the largest house in the area. By the way, how do you manage to survive in these mountains"

"We have a few goats and plant vegetables. Every male here is an excellent marksman and we hunt for the wealthy padrones."

"Do they get paid for that?"

Mario's Trek To Pico

"Of course we get paid, but the war has interrupted business."

They entered the house which was very clean and had an earthen floor. His wife, a portly woman was praying near her altar.

Antonio placed a finger near his nose, "Shhh, she is praying for herself and me."

"Don't you help her at all?" Mario kidded.

"No, it's not necessary, it is her duty and, so far, it has worked. We are both safe and have food. God is good."

She rose and turned as Antonio said, "Look, Maria, an American."

She came to Mario and embraced him, "Tonight, we will celebrate by sharing our food with you." Mario kissed her on both cheeks and she left to prepare a small dinner.

Antonio made himself comfortable in a small chair and directed his guest to do the same. Mario removed his combat gear and sipped on a glass of wine. He asked, "How does that hunting work regarding the padrones?"

"The most important thing is that the game cannot be destroyed. One must be proficient by shooting the prey in the head only."

"No kidding, that must be difficult."

He patted Mario on the knee. "Not from someone who is born and raised in this village. We are the best hunters in all of Italy."

"How about the descendants who now reside in America?"

Antonio replied matter of factly, "It doesn't matter where they live, it's in the blood. Why do you ask so many questions, you sure are nosey."

Mario liked the old man. Maria brought some goat cheese and bread to the small table. She apologized for the stale condition of the bread. "It is too dangerous to make a fire for baking. Now that you Americans are here we can bake again."

"No," replied Mario, "I'm alone." He asked the old couple, "Have you ever heard of the Di Lucci name?"

They beamed at the mention of the name. Antonio said, "That family has been here for as far as I can remember. We are cousins.

"Do you remember three Di Lucci brothers by the names of Giovanni, Antonio and Salvatore. They went to America as very young men."

"Yes, I remember them well, especially Salvatore. They all shot well, but he was the best in this village."

"Giovanni is my grandfather and, of course, Salvatore and Antonio are my granduncles."

Pico

Maria began to cry with joy and embraced Mario. She kept repeating, "Our prayers are answered, our prayers are answered."

Antonio soon had tears in his eyes, "You are a distant cousin. The Almighty looks down favorably on us."

Mario was happy to arrive at the top of a little hamlet and find some of his roots. The old peasant offered to round up the villagers for a celebration. Mario respectfully declined explaining about his objective.

Antonio informed him with deep concern, "There was a lot of shooting there a few days ago. Pico is very important to the Germans and the front lines keep changing there. I don't want to alarm you, but you should stay here awhile."

"Do you ever go to Pico?" Mario asked.

"Yes, that is the big city for us. We used to go there to trade cheese for grain and occasionally sell wood, but since the war intensified we couldn't go anymore."

"Do you know any people there?"

"Of course. We are nearly paisanos."

"Do you know the Calcagnis?"

He smiled with delight, "Everybody knows the Calcagnis, they run the place."

"My mother is a DiLucci and she married my father, Rocco Calcagni." They embraced again and related to him that they were aware of the wedding. Maria said, "They had one son."

"You're looking at him," Mario said.

They now looked upon him as their adopted son and Maria ordered, "You must stay here for at least a week."

Mario thanked her for her concern and laughed, "My superiors in the army would throw me in jail if I took you up on your generous offer but I've got to leave in the morning." He clasped his hands and pleaded with a smile, "You both must excuse me, I need to get some sleep."

He laid on a straw bed on the floor as Maria covered him with a blanket and said, "Sleep, my son, sleep well."

Antonio told Maria, "It sure is a small world, isn't it?"

She went to her altar and prayed for Mario.

The next morning, Mario awakened before dawn. Maria heard him stirring and was up before him. She woke up Antonio and they all had some stale bread with goat milk.

Mario's Trek To Pico

"I noticed a horse tied in the far end of your corral," Mario said.

"It belongs to the Germans. They kept the horses when they rode into the hills. This one was left behind. You know, Mario, speaking of the Calcagnis, I saw them in the mountains about a month ago. In fact, they could very well be hiding nearby."

"I wish I could run into my uncles somewhere around here."

"It is strongly rumored that your uncle Giovanni killed two Germans and they were looking for him. There are many rumors during hard times but I believe he did it. The brothers are heroes in the area."

"Do you mind if I take the horse?"

"Take her and don't worry about it. The Germans are gone and a horse is useless in this area. Only mules and jackasses count."

Mario gathered up the saddle and got acquainted with the mare. He asked Antonio if there was an easier trail to the main road. The American embraced them and thanked them for the food and a good night's sleep. He mounted the horse as the old man pointed out the direction and told him to turn right at the fork as it would be a safer trail.

Antonio grabbed his hand one more time and asked, "I've been thinking all night long, are you a good shot?"

Mario smiled with confidence, "I'm the best there is!"

The old man tugged and snapped his suspenders and announced happily, "I knew it."

It was approximately nine in the morning when Mario reached the road to Pico. There was shooting in the mountains, but the road was quiet.

He thought: Perhaps the lull before the storm. He planned on traveling the last three or four miles at breakneck speed. The idea was to arrive at the outskirt of his destination in a matter of fifteen minutes or so. Mario was so close, he could feel it. After removing the white scarf from his pocket, he wrapped it around his neck. The horse reared on its two hind legs which caused him to smile. Mario patted the mare on the neck and spoke to her, "You're full of run, I wish I knew your name."

The horse responded from the gentle tapping as her ears perked up. He had enough experience to realize that the animal hadn't been exercised for the past month. The sergeant felt the same way.

Pico

He hadn't been on a horse for nearly a year and thought, this sure feels good.

The air was crisp on this spring morning. He observed the countless bushes of yellow flowers gracing the side of the nearby hills and contemplated: It's a lovely day to be alive. If only this were peacetime instead of war. He looked around and was surprised that the road section was eerily quiet. From evidence observed in the little hamlet, Mario surmised that the Moroccan Goumiers outflanked the Germans somewhere in the immediate area. It could have happened a few days ago but there still existed pockets of enemy resistance. Some Germans were ordered to hold their occupied positions to death and they obeyed. He understood the risk involved by being on horseback but had a weird sensation that time was of the essence. It is now or never. He reared the spirited mare and slapped her hard on the flanks and cut her loose. Mario was on his way.

The road was rough in certain sections but the mare adjusted accordingly. She jumped when she had to and Mario guided her superbly through obstacles of destroyed tanks and debris of metal. They raced through some smoldering fires which emitted a substantial amount of acrid smoke. The mare was oblivious to the surrounding conditions. Mario deduced that the Germans did a great job in training this animal.

He spoke to her as if she understood, "Girl, I wish I could bring you home with me." Unknown to Mario, the Hitler line had faltered and was readjusted to reform at a new position known as the Dora line. At best, it contained sporadic pockets of defenders who were in a state of disorientation. The line was formed as a last-ditch effort to hold the supply base at Pico. Mario was under the impression that he had a reasonable chance for a clear road ahead. He knew that there was only a remote chance that Goumiers would shoot at him. The sergeant was within two miles from the outskirts of his objective when he was observed by a small squad of Germans. Horst was in charge of an anti-tank gun emplacement that was setup to protect Pico from the west. He didn't realize it but the Goumiers had crossed the road and penetrated into Pastena. Horst had an excellent position but had very little support from the mountains to the north. Units of his company had retreated back toward San Giovanni to form another defense position.

Mario's Trek To Pico

Horst and his group were alone but they didn't realize it. They were ordered to stop oncoming armor at any cost and had the capability but in reality, were in no man's land. Horst was looking eastward on the road and scanning the nearby mountains with binoculars as one of his men tapped him on the shoulder. "Horst, turn around and look what's heading towards us!"

He turned and focused his binoculars. He uttered in amazement, "I don't believe it!" The rider was skillfully weaving through the destroyed tanks. The smoke obstructed their view momentarily. Mario and his mare broke into the clear and began gaining momentum. Horse and rider were approximately four hundred yards away from the their position and closing fast. The rider was positioned like a jockey with his upper body shielded by the horse's neck.

His white scarf was flowing with the wind as Horst refocused the glasses. Horst ordered his rifleman to focus on the rider but instructed him to wait for the order to shoot. He raised his left hand as the rider was less than two hundred yards away. "I have him dead in my sights; just say when." Horst was hesitating the command.

There was something odd about the entire situation. The rider was within one hundred yards and had clear sailing. The Germans were well hidden and Mario was unaware of what was ahead. Horst knew he had enough time to delay a decision of whether to shoot the rider or not. He spoke out, "This guy has a lot of balls. I would like to meet him." He then noticed that the rider was an American and thought, "What the hell is an American doing here?" Horst had his hand held high and gave the order, "Take out the horse!"

Mario was within one hundred feet when the shot rang out. The animal's legs buckled as Mario went flying over its head and crashed in a ditch. He landed within thirty feet of the Germans. Mario was temporarily stunned which allowed his capture. He was still dazed when they dragged him into their position.

Horst looked at Mario's scarf and noticed the name Calcagni. Mario was semiconscious as one of the Germans splashed some water on his face. Mario came around and drank some water. He felt around his body and was satisfied that there were no broken bones. Bruised and hurting, he had survived the nasty fall. Horst asked him in Italian, "What are you, doing on this road?"

Pico

Mario looked at him and replied with his classical phrase, "Fuck you."

Horst sensed something in this adversary that interested him with each passing moment. He thought: Why the scarf, not many soldiers are allowed to be out of uniform. In addition, the name Calcagni was very familiar.

Mario was contemplating in retrieving his weapon. He reasoned that as soon as the opportunity came he would hit this lax German who was questioning him and figured that he had an even chance with the others. Two of them were wounded and not at full physical capacity.

"I believe your name is Calcagni. I knew people in Pico by that name."

Mario's eyes brightened, "I have uncles there as well as a grandmother and countless cousins." He thought, with renewed hope, luck is sure with me, here lies the enemy and he appears to know my relatives.

"I used to call your grandmother, Mama Maria."

Mario became at ease and took his eyes off his weapon. He answered, "Yes, my grandmother's name is Maria."

"Your uncles are a little on the wild side but they are very good friends with the Micheletti family."

"I have never heard of that family. What interest do you have with an Italian family?"

Horst replied confidently, "After this war is over, I am going to marry Gabriella Micheletti and live on their farm."

Mario walked over to his weapon and retrieved it. One of the Germans leveled his rifle at Mario but was waved off by Horst. Mario asked him, "What's your name?"

"Horst Schultz."

"Sergeant Schultz, the war is over for you. The situation here is hopeless, you and your men are alone."

"You don't understand the German Army. We have our orders and even if we wanted to leave, the SS is not far behind. They are ordered to shoot whoever leaves their position."

Mario was checking his weapon and grenades and took a moment to look at Horst, "Fuck the SS." The German was perplexed by the remark as Mario continued, "Yesterday, I killed six of those fucking Moroccan Goumiers. I fear that they are committing

Mario's Trek To Pico

atrocities in Pico. These people are animals and will rape anything in sight." Mario shook Horst's hand and asked, "Why don't you come with me?"

The German forced a laugh, "What chance do I have with the enemy? Pico is already in their hands. The Goumiers will shoot me because I will never allow them to take me prisoner. As bad as they are, you are all on the same side which will enable you to get the drop on them." Horst sat down and covered his face with his hands, "Gabriella, my sweet Gabriella, I am so close and yet so distant. I pray that you have somehow escaped the perils of war."

Mario sat next to him and advised, "It's all over for you, forget about the SS and head down this road towards Itri. Give yourselves up to the Americans there. There will be no SS in that direction. If you run into straggling Goumiers, shoot the bastards."

Horst asked his men, "What do you guys think? Shall we make for the American sector and surrender to them?"

They looked at each other, then at Horst and finally Mario, who assured them with a nod. They rose to their feet as Mario asked, "What is the best way to get to Pico?"

Horst pointed down the road to the northeast. "You go up the road there near that big rock in the curve. The trail starts in front of that rock. I heard that the going is treacherous but it will bring you near the Micheletti property."

"How far is the trail from the road?"

"I have never used it but the distance must be close to two meters."

Mario asked out of curiosity, "You were so close, why didn't you chance it?"

Horst led Mario out of earshot of the other soldiers. He explained, "I wanted to defect from the army and hide in the hills with the Michelettis but they refused."

"Why would they refuse?"

"I find that Italian families are very close. The mother, Anna, wouldn't leave her parents and they had refused to go."

"That was very thoughtful of you but why didn't the old folks consent?"

"You wouldn't believe it, but the old man said he had to make wine. For all I know, they could all be dead and all over some wine."

Pico

"You may have had a problem hiding a family in the high mountains."

"No problem. I had plans to tag along with your uncles. Besides, Gabriella has a younger sister called Angelina who knows the mountains like the back of her hand. She is a little spitfire."

Mario smiled, "I'll look for her if I get lost." He turned to leave as Horst requested, "Please look up Gabriella for me and tell her that I'm safe and that I love her."

Mario answered, "I will give your message and tell her that you are with the Americans."

Horst watched and gave Mario cover as he entered the trail. The Germans gathered their equipment and proceeded west towards Itri. Mario looked back and saw Horst standing tall with an automatic weapon in his hands. He gave a final wave and said under his breath: Nice fucking guy. I wouldn't mind having him at my side during a battle.

The trail was difficult. The constant rains had caused minor landslides that forced him to drop down, around and up again. In other cases, he went up, around and down again. He had to be alert because his enemy was now the Goumiers and he didn't want to be surprised by them. The Germans were gone. There was very little shooting and the heavy shelling was in an arc that included Pastena, San Giovanni, Ceprano and Arce. There were times that he heard movement and stopped as a precautionary measure.

Whenever he did see anyone, they were civilians who were afraid and trying to escape from inhabited areas.

Mario was thinking about Horst and then smiled to himself: He shot my fucking horse.

20

PICO ARRIVAL

As Mario reached the summit of the ridge he saw Pico for the first time. It wasn't a pleasant sight. There were numerous columns of smoke arising from sections of the town. It was as beautiful as any little mountain village in Italy. He couldn't determine the extent of the activity, but reasoned that the Germans had finally retreated. The buildings that he could distinguish had tile roofs. Most of them had architectural designs, but all were painted the same tan color. The streets were obstructed by buildings that were built close together.

The highest buildings were the churches with their ever present steeples. He came upon a wide trail and estimated that Pico was perhaps fifteen minutes away. Mario realized the only danger to him would be the Moroccan Goumiers. He made up his mind that if a confrontation with them arose, he was prepared to shoot first and let the chips fall where they may. The chances of meeting them on the main trail were high. He had his weapon set in a firing mode. Mario had proceeded approximately twenty meters when he heard muffled noises ahead. He stopped and took cover. A group of peasants approached. They looked tired, shabby and malnourished. It was the first time that Mario was ready to face his fellow Picanos. Mario stepped out into the trail and halted their progress. The natives were startled and crowded together.

Mario pointed his weapon upward and spoke in Italian, "I'm an American." They became jubilant as the occasion allowed. He waved and explained, "Hold on, I'm alone. I am only one American

Pico

within fifteen miles of here." They responded, "God bless you, Americano, we are saved."

"Where are you people going?"

Some of them tried to speak at the same time and Mario became confused. He held up a hand, "One at a time and don't talk so fast. I am good with the language but not that good."

A young girl spoke out, "The Goumiers are in the area and we are trying to hide from them."

Mario already knew, "Why?"

One by one, they all gave different stories of atrocities committed on them and other Italians. The attacks on their persons varied but the results were all the same. Rape, beatings, injury and death. They wept as they related their experiences and most of them couldn't complete a sentence without breaking down in grief. The girl stepped forward and said, "You must drive these devils from this community."

Mario looked at her curiously and asked, "What is your name?"

"Angelina. What is your name?"

"Mario Calcagni and I'm looking for my family."

Angelina's eyes lit up. She was experiencing a feeling that was alien to her. It was love at first sight. Angelina was, for the first time, speechless.

Mario's instincts commanded him to continuously scan the surroundings as he obviously did not notice the reaction from the young Italian girl. He noticed movement up ahead and prepared himself without alarming the group. Mario felt they could be Goumiers. A few in the group heard the same noises and became frightened.

Suddenly, four Goumiers appeared and were running towards the civilians. They had long knives drawn and squealed to terrify them.

Mario shoved Angelina to the ground and ordered everyone to lay down. He leveled his weapon to a firing position as the Goumiers stopped in their tracks. They had knives drawn but their guns were on their shoulders.

One of the Goumiers began speaking French and the only thing that Mario could determine was that the civilians were nothing more than spoils of war. Mario smiled, in disgust. They obviously misunderstood the smile and lowered their arms. They cautiously

Pico Arrival

approached as Mario fired. The barbarians dropped to the ground as Mario closed in on them and fired again. The smoke soon cleared and the peasants began rising. Mario turned and ordered them to stay down. He walked ahead around the trail and spotted three more Goumiers who were running away. Mario fired and shot all three. He moved forward and inspected the inert bodies. There was no need to expend more ammunition, they were dead.

Angelina came running, he turned to her and said, "It is still dangerous here. Stay put."

She looked him defiantly, "I'm not afraid of anything in this world."

Mario scanned the area while thinking of a suitable reply. There was something about this Angelina, but he couldn't put his finger on it. "Do you have a sister named Gabriella?"

Angelina was dumbfounded, "What could you possibly know about my sister?"

"I met her German boyfriend."

"You mean Horst, is he still alive?" Angelina felt that any soldier who comes into contact with Mario has a good chance of being dead.

"Yes, he's alive, and by now could well be safely in American-occupied territory." Angelina gave a sigh of relief. Mario waved the group to come ahead. He ordered the men to hide the dead Goumiers in the brush. He also told them to gather up all the weapons.

Angelina picked up a rifle. Mario laughed, "What are you going to do with that?"

"I know how to use it."

Mario shrugged and asked, "Does anyone here know my family?"

"Who are you?"

The American looked into her eyes and replied with a smile, "It isn't any of your business."

"Everybody knows the Calcagnis. They are our best friends."

"Where are they?"

"Your two uncles are hiding in the hills somewhere with their family," Anna replied.

"Is my grandmother with them?"

"No. She is not well enough to travel. We have her hidden in the attic of my house." Anna had a troubled, sad look on her face.

Pico

Mario noticed the expression and wanted more information about the old lady. Anna was trying to avoid the conversation but Angelina was ready to speak. Anna gave her daughter an adminishing look that Mario comprehended and became suspiciously concerned.

"What is wrong with my grandmother?" he asked again.

Anna skirted the question and replied, "Your uncles, Giovanni and Luigi killed two of the Goumiers and hid into the mountains."

"Why did they do that?"

Anna didn't answer but an old woman was ready to offer the rest when someone nudged her. Mario caught the gesture and asked pointedly and a bit irritated, "For the last time and I want an answer. What's wrong with my grandmother?"

A younger woman replied over the stillness, "They raped your grandmother."

"She is nearly eighty years old. Are they crazy?"

"No, my son, they raped her six times and left her for dead."

"Is she alive?"

"Yes, but she may as well be dead, the poor soul."

Mario bowed his head momentarily. They were quiet. He looked up and asked with a quiver in his voice, "Does anyone know where my uncles are?"

"Yes, I know where they are and I can find them."

Mario gazed over at the girl somewhat puzzled, "It's dangerous and you're too young to show me the way."

"I can do just about anything you can do."

Mario would have burst out laughing if it wasn't for the news concerning his grandmother.

"She knows the mountains better than anyone and she can elude the devil himself," Anna said.

"Then you must have a great deal of faith in this little girl?"

"I'm not a little girl."

Anna gave her another one of her piercing looks but knew that her Angelina was in love, "Yes, I do, Angelina is very special."

Mario smiled, "I heard the same comments from the German. By the way, where is Gabriella?" She was standing near her mother throughout the conversation but remained quiet. Her shyness was the reason why Mario didn't notice her. Anna introduced her and she bowed her head and said hello. Angelina offered an explanation that Gabriella doesn't talk much until she becomes comfortable with strangers. Mario said, "I'm not a stranger."

Pico Arrival

Angelina quickly replied, "I know."

"Does anyone here know how many of these Moroccans are around here?"

They all tried to answer at the same time. Some said fifty. Others guessed one hundred and still others estimated more. Anna said, "I think there were a lot of them but it appears that they moved on. The problem is that they keep coming back."

Mario tried to evaluate the situation. "Is it possible that they are not supposed to be here?" Angelina was listening and told what she had recently seen. She saw a French officer shoot a Goumier. The French were ordering them out of town but the Goumiers defied.

Mario knew that the French commanders created a situation that got out of control. He rationalized that the forward units voluntarily stopped in the mountains north of Pastena and were taking unauthorized leave by returning to Pico and picking up where they left off. If this were true, he determined he could clean them out without much problem.

Angelina was staring at him and asked, "What are you thinking about?"

"It looks like you and I are going to look for my uncles."

She brightened and swung the rifle over her shoulder. Mario shook his head. "Little Angelina, in about ten minutes, that weapon is going to weigh a ton."

"Don't call me little."

Anna told Gabriella, "It's unfortunate that Angelina is so young. They would make an excellent couple."

Gabriella nudged her mother, "You don't know Angelina like I do, Mario is already hooked but doesn't know it."

The American assembled the group and explained his plan. "I want everybody to go to the farmhouse where my grandmother is. Stay in the house and be quiet. The men will be on guard with the weapons they now have. If the Moroccans return, shoot them. Don't worry about reprisals, let me worry about that. I'll be back with my uncles in a couple of hours." He looked at Angelina for confirmation. She nodded in agreement. He instructed the men on the use of the weapons. He also told the women to pay attention.

Departing, Mario informed them, the situation has changed. "I believe that these bastards are being hunted by their officers.

Pico

Whatever you do will be in self-defense. When I return, I'll take care of business." They appeared to understand as he and Angelina prepared to leave.

It was early afternoon when they left and headed in the same direction that Mario originally came from. Mario soon stopped and pulled out his map. He sat on a rock and pointed out their present position and asked Angelina to show him their destination.

She looked and scoffed, "I don't care about that stupid piece of paper."

"It doesn't matter whether you care or not, just show me the direction. After we go around one hill, we may be heading south."

She waved her hands in the air, "The Calcagni brothers are hiding in a cave near Campodimele."

"I think I just came from there."

Angelina changed the subject. "Your uncle Luigi is a fine singer. Do you sing?"

"Yes, I can sing, how about you?"

She broke into a big smile, "Yes, and I love to sing in the mountains. A voice echoing off the mountains is the most beautiful sound in the world."

"It's too bad that the war is here all around us or we could sing a few choruses," Mario mused.

"Do you know Santa Lucia?" Angelina probed.

Yes, of course I know that song, it's one of my favorites."

"Let's sing it."

"Are you crazy? Those Goumiers will be all around us."

She aimed her rifle and looked at Mario, "Too bad for them, they wouldn't have a chance."

Mario shook his head and pointed to the trail, "Let's go."

She moved along effortlessly. Mario couldn't believe the energy and agility displayed by the little girl. Angelina negotiated the inclines as if they were level paths.

He was in good physical condition but had difficulty in keeping up with her. After about fifteen minutes, he signaled her to stop which gave him the opportunity to catch up. Angelina said, "What's the matter, Mario, can't you handle the mountains?"

He smiled faintly, hiding his fatigue, "I was born and raised on flat land. I'm not as good as you up here, but I can manage." He then winked at her, "You are an exception."

Pico Arrival

The birds were singing which put him at ease. She noticed that he was giving a great deal of attention to crows in the distant skies. Their presence in the area, except from the general location occupied by Mario and Angelina, indicated there was no one around. He recalled his youth and remembered how difficult it was to get a shot at a crow. There were so many of them but the black birds always managed to stay out of range. The farmers considered the scavengers detrimental to their crop. The birds ate the seeds the farmers planted. Angelina broke his daydreaming, "Why are you looking at the crows? Are you going to shoot at them?"

"No, little one. I'm just observing. Those black scavengers are lucky for me."

"Don't call me 'little one' and besides, crows are good for nothing."

"They are helping us today. Their eyes and senses are telling us that it is safe ahead."

She thought a moment and replied, "Mario, you sure are smart. The Almighty has been very good to you."

"Thanks. Tell me about your sister Gabriella. Why is she so quiet?"

Angelina thought for awhile and knew that, sooner or later, he would find out all about the tragedies suffered by her sister. "Gabriella is my cousin, but my mother raised her since she was a baby, so that makes her my sister. I love her as much as I love my mother." Angelina related the events of the fattura.

Mario was saddened and revealed that he heard about the pagan ritual. He said he never believed in witchcraft which only occurred in primitive areas where the innocents are fed a dose of poison. He added that he believed very little of what he didn't see with his own eyes.

"My sister is the most beautiful girl in Pico."

"Yes, I agree but remember, beauty isn't everything." He noticed she became somewhat disappointed. Mario added, "Angelina, you are a beautiful girl, when you become a young lady in a few years, you will be competing with the goddess Venus."

Her eyes brightened, "Do you really think so?"

"I assure you, I am a judge of beauty."

"Mario, what makes you think that I am going to be beautiful?"

Pico

He took a deep sigh, "Do you consider your mother beautiful?"

"Yes, of course, when she was young, she was a Venus."

"You resemble her to a tee and will look like her when you grow up."

She looked at him with adoring eyes and said, "You know Mario, you are not such a bad person." She became serious, "Horst killed the two people who were involved with the fattura. Gabriella nearly died and he saved her life."

"From what contact I had with him, Horst indicated he loves her very much."

"I know that for a fact but he may change his mind when he discovers what happened to poor Gabriella."

She rose to get moving again but Mario stopped her, "What is wrong?"

Angelina bowed her head as tears came to her eyes. He lifted her chin with his hand and tried to comfort her. She replied between sobbing, "The Goumiers beat and raped her. If Horst finds out, he won't want to ever see her again."

Mario replied softly, "Any man who will kill for a woman he loves will overcome any problems. He will destroy whatever barriers confront him. Angelina, love is so powerful that it cannot be subdued."

"What would you do if you were in his position?"

"I have never been that much in love but I know how I would respond."

Her face glowed for two reasons. She knew exactly how he would face the problem and was elated to know that this Calcagni of Calcagnis had never been truly in real love.

He didn't comprehend the expression on her face and said, "May I ask you a personal question?"

She looked at him curiously, "Yes."

Mario hesitated, he was conversing with a child and the question was of a delicate nature.

She egged him on, "What's on your mind?"

"Did the Goumiers take advantage of you?"

Angelina was enraged, "No, never! But they tried. Come on, let's move on."

Mario shrugged and thought to himself: I should have minded my own business.

Pico Arrival

She moved even faster as the sergeant followed her across a steep ravine which they climbed to the summit of a ridge when shots rang out, momentarily halting their progress. Finding a secure position, they soon had a clear view of the action. Angelina explained, "Your uncles are in that cave on the other side of the ravine."

Mario scanned the area and observed that a squad of Goumiers had the cave well covered and were directing an assault into the mouth of the cave. He needed to respond quickly. Mario asked Angelina if she had ever been in the cave. She nodded and described the layout. The cave opening went in about thirty feet and veered off to the left for an additional twenty feet. The American determined the family was relatively safe for the time being. He also realized why the Goumiers didn't close in. His uncles had weapons and were returning the fire. It wasn't much but they had an advantageous position.

Mario quickly evaluated the situation and determined his plan of attack. He could account for ten of his adversaries that were closely bunched. It was a mistake on their part but they were so sure of themselves because of the weak resistance. The Goumiers were patient and playing out a game. It was understandable, they didn't want to kill all the occupants. The prime objective was rape.

The resistance from the cave was feeble and Mario feared his uncles were getting low on ammunition. The entrance to the cave was approximately three hundred feet away. The Goumiers were a few hundred feet away from his position. He needed to get closer so he could gain an advantage. Grenades, submachine gun and the all-important element of surprise. He turned to Angelina, "Keep you eyes open and cover me."

She tried to ignore him but to no avail. He expressed sternly, "I don't have the time to argue with you. Stay here and do as I tell you." She looked at him momentarily and finally agreed. As he was leaving, he pointed a menacing finger in her direction and disappeared stealthily in the brush.

Angelina observed with anticipation and awaited the sight of her sergeant. The Goumiers had no reason to believe that anyone was behind them. The Germans were gone. Their only concern was their French officers who were back in Pico.

Mario suddenly appeared and positioned himself on a flat rock. After a quick assessment of his overall position, he threw four grenades in rapid succession. There was no return fire, he caught them completely by surprise. The grenades found their mark and he opened fire with his automatic weapon for insurance. He pressumed they were all dead.

Angelina saw one enemy escaping from a point that was not visible to Mario. She fired the weapon and missed. Mario turned around and realized what happened. He waved her down and she explained that one of them got away. Mario was distressed to hear this. The Goumiers would soon be alerted and ready. He told her, "This is the first one to get away since I started on this trip."

Unknown to the American, two others were hidden in the rocks to the right. They returned a fusillade of firepower that went over Mario's head and the direction of Angelina. He dove for cover and suddenly realized that she was in danger.

Mario was not one to panic, especially in combat situations. He yelled out, "Angelina!" There was no reply.

The firing continued as Mario determined the enemy to be less than one hundred feet from his pinned down position. This time, desperate with concern, he yelled, "Angelina!" There was no response. "Son of a bitch, they shot her." He readied two grenades and determined the position of the well fortified Goumiers. His mind flashed to the time when he practiced one summer throwing the javelin. He pulled the pins and arched each grenade in a high trajectory.

They found their mark behind the large boulders. The explosion was followed by silence as Mario rushed to the position. The barbarians were dead but he made sure they were by pumping bullets into them. Mario composed himself as his thoughts went back to Angelina. He ran with all possible speed he could muster screaming her name.

He stopped at the large rock that he last saw her upon and looked down to a twenty-foot drop. The girl was inert. Mario made his way down expecting the worst. He placed his weapon on the ground and tenderly lifted her limp body.

Mario was saddened and cried, feeling responsible for her. "Angelina, my poor little Angelina."

He checked her over and didn't notice any sign of bullet wounds. Mario gently rubbed her forehead as she stirred. Her eyes opened but she was having difficulty trying to talk. Mario smiled with a

Pico Arrival

sigh of relief, "Thank God, you're okay. The fall knocked the wind out of you."

Angelina spoke softly, "The bullets were so close that I jumped and landed on my shoulder. I think I tripped when I started out."

Mario embraced her as she massaged her shoulder, "You scared the hell out of me, I thought you were hit." Satisfied that she was okay, he helped her to rise, "You'll be fine in a few minutes." Angelina smiled and felt that Mario had a special place in his heart for her.

Mario placed his weapon over his shoulder, "Pick up your rifle and let's go meet my family."

"Yes sir!" she joked and feigned a limp.

"What's wrong with you now?"

"My leg hurts, I can't walk."

"I suppose you want me to carry you?"

"Well, Sergeant, I can't walk."

He lifted her to a nearby rock, turned and offered his back. Angelina kicked him playfully in the rear and climbed on his back. He shook his head, "In America, we call it piggyback." Angelina rested her head comfortably on the side of his neck. Mario asked curiously, "I suppose that your head also hurts."

"It feels fine now."

He smiled with raised eyebrows, "Call out to my uncles before they start shooting at us."

"Don't worry, I have a better idea, I'll sing and they will know who I am."

"How well do you know my uncles?"

"I know them as if they were my immediate family. They are my heroes. The Calcagnis are real people." She cuddled her head closer, "Mario."

"Now what?"

Angelina hugged him tightly, "I love you!"

He hesitated for a few moments and replied softly as he put her down, "Angelina, you're only a child and think you love me. Trust me, this will all soon be a memory. Let's move, I know that your leg is better now."

"Mario, one more thing."

"What else is on that inquisitive mind?"

"Would you love me if I were five years older?"

"Yes."

21

THE CALCAGNIS MEET AND CLEAR THE AREA OF THE ENEMY

The occupants in the cave heard the fiery exchange outside. Giovanni and Luigi were lying prone on the floor. Giovanni looked up and said, "I think those Moroccan devils were ambushed, but by whom?"

"It's probably the Germans, they keep counterattacking," Luigi replied.

They knew they were in trouble regardless of who was out there. Everybody was the enemy but their chances were better with the Germans. The Calcagni brothers didn't hear any movement but they knew someone was coming. They aimed their rifles towards the trail when they heard a voice singing, "Santa Lucia."

Giovanni jumped up and shouted, "Angelina!"

He helped his brother to his feet. Luigi had taken a bullet in the leg. When the rest of the family came forth and joined Giovanni and Luigi, they looked out and saw the American with Angelina.

"We have a surprise for you," she said.

They stared at the soldier and thought they recognized him under the helmet and a three-day beard. He spoke, "I am the son of your brother, Rocco."

The women reacted in unison, "Oh, my God!" Giovanni said. Luigi remained speechless.

"Your nephew, Mario, killed them all except one that was apart from the others and he got away," Angelina said.

The two brothers approached Mario and took turns embracing and kissing him on each cheek. The women and children were understandably shy.

The Calcagnis Meet and Clear The Area of The Enemy

"It took our nephew all the way from America to come here and save the lives of our family," Giovanni said.

Mario took Angelina's hand and said, "She deserves the credit. Without her, I would never have found you."

"Yes, Angelina, this brave and compassionate girl is our favorite person," Giovanni added.

After a perfunctory talk about the family, Mario informed them that the Goumier who escaped could be back with reinforcements and the element of surprise will not work. They agreed and decided they should all go to the Micheletti farm.

The women tended Luigi's wound and found a sturdy branch of a tree and improvised a crutch. The brothers selected the best weapons from the dead and supplied themselves with grenades. Mario was pleased. He now had better position and strength.

Mario explained that the Goumiers were acting on their own. They were A.W.O.L. and were not about to tell their superiors about what happened. He felt that the three of them could go on the offense and help rid the devils out of the area.

Even though it was slow going with the hobbling Luigi, they reached the farmhouse as darkness set in. The occupants of the house saw them coming and ran out to meet them. They were happy and felt a sense of security. The men had become a threat to the barbarians.

"We have a very small problem. About an hour ago, several people came here and forced their way into the house. One of them has a pistol," Anna said.

"Who are they?" Mario asked.

"They are local politicians who collaborated with the Germans and now they are afraid of the Moroccans. They are no good and everybody despises them.

"Show me who they are!" Mario ordered.

"They have food, but won't share it. Be careful," Anna said.

The collaborators were the only ones remaining in the house as Mario entered with his sub-machine gun and leveled it at them. He asked, "Who in the fuck are you people?" There was no reply.

Anna pointed out the leader. Mario approached him and asked him for his pistol. He denied he had a weapon. Mario slapped him with an open hand that drove him across the room. One of his cohorts tried to interfere and Mario punched him and knocked him unconscious. The Calcagni brothers were ready if they were needed.

Mario handed his weapon to Angelina and harassed the collaborators. The first one never went for his pistol. Mario picked him up by the neck punched him again.

Mario took his gun and ordered them all out. The women taking the food along with them. He stopped them, "The food stays here. Now get the fuck out of here before I shoot the whole bunch of you."

One woman spat at Mario. He responded by kicking her in the ass. Everybody laughed. Mario said it was the first time he had ever hit a woman. Luigi replied, "You did good nephew, if you didn't do it, I would have, bad leg and all." They all returned into the house and the women set a table with the confiscated food.

They were festive for the first time in nearly a year. Giovanni told his brother, "It sure makes a difference with Mario here." Luigi said, "The kid's a one-man army." Mario posted the most alert of the women to keep a watchful eye for Goumiers. The others sat around and discussed the possible ways to bring back normalcy.

The consensus was that the Goumiers were permitted to commit atrocities for a period of six days or so. They heard rumors that it was designated as the White Paper Act but no one could verify this. Mario was now trying to establish where this authorization came from. The only thing he knew however, was that the atrocities occurred and everyone had an experience to relate. There was enough evidence that the army lost control and could no longer stop the Goumiers. They were running wild, seeking whatever they desired and were successful so far.

Mario was mildly concerned with the Goumier who escaped the cave. He expected only a token reprisal from what the barbarian could convince of his cohorts. Yet, there was the other view that they were afraid of the American. They would never disclose what actually happened to the civilians. Mario conveyed his thoughts to his uncles and they agreed. He added that up until now, the dead was being blamed on the Germans and he intended to keep it that way. It was possible that the Goumier didn't see him.

"Do you think the one who got away knew that I am an American?" he asked Angelina.

"It's hard to say because he ran away fast," she replied.

The Calcagnis Meet and Clear The Area of The Enemy

Giovanni excused himself and went up into the attic to visit his mother. He brought her some wine and food. Anna was attending to Luigi's leg. The bullet passed completely through. She dressed the wound with some salve that Mario provided from his first-aid kit. Giovanni came back and Mario asked to see his grandmother.

"You can if you wish, but she doesn't recognize anyone. My mother is in shock, I don't believe that she will ever recover."

"I still need to see her, both for myself and for my father."

Giovanni sadly pointed the way. Mario climbed the ladder and saw an old woman coiled up like a fetus. She was trembling and had her face covered with a kerchief. Mario approached her as Giovanni watched from the top of the ladder. "Mama, Mama, it's me, Mario. I am the son of Rocco." She was sobbing and the sound of the name Rocco appeared to faintly awaken her from her world of oblivion. Mario cuddled her in his massive arms and sang softly the words to "Santa Lucia." Giovanni observed with keen interest. He signaled Anita to come up the ladder as he quietly moved onto the small loft. It appeared that a miracle was happening, she thought she recognized Mario as her son.

The old lady put her arms around Mario and cried softly, "My Rocco, you have finally come home to your mother."

He held her in his arms as she slipped into sleep. Her trembling stopped and she relaxed with a trace of a smile.

"It's a miracle, Mama is going to recover," Anita announced to all.

Giovanni embraced his nephew and asked, "What more can you do for our family?"

"It isn't over yet, we are all in danger and must stay alert."

"That is the truth," Luigi replied.

About ten in the evening they were alerted that someone was approaching the house. Giovanni looked out a window and saw two Goumiers who were within a few feet of the door and began their familiar squeal. They banged on the door with their rifle butts and Anna told them to go away. More of the women cried out the same. The Goumiers stopped as Mario signaled to Anna to open the door and the Moroccans rushed in expecting no resistance.

Mario grabbed both of them at the same time and threw them out the door. They landed on each other and Mario began kicking them. He had them where he wanted them. Their weapons were scattered as he began beating them relentlessly.

Pico

Giovanni and Luigi had guns pointed at them as Mario picked one of them up and scowled, "Let me hear your fucking squeal now." He was too terrified to answer so Mario beat him into submission.

The other Goumier cowered as Mario said, "You're next." He belted him in the face until it became a mass of broken bones and blood. Mario showed the Italians that the barbarians were nothing but cowards. It was as if to help erase the atrocities that the onlookers were subjected to. Mario wiped his hands with a cloth and cussed under his breath, "Fucking no-good bastards." He threw the bloody cloth on the ground and asked a few of the old-timers to drag the Goumiers up to the barn. They were happy to drag them by their feet.

The men entered the barn. Giovanni threw a pail of water on their face. The elders held them up by the arms as Giovanni paced back and forth in front of them with a long knife flashing at them. Mario ordered Angelina to leave but she refused.

Someone noticed a Goumier dead on the earthen floor. Another bystander felt a drop of liquid on his bald head.

"What the fuck is this dead Goumier doing there?" Mario said.

"There is another one dead up in the loft," Angelina said.

Mario threw up his head, "How do you know there is a dead one up there?"

"Because I shot them both."

Mario scratched his head, "Angelina, will you please leave?" She left but positioned herself near the barn door.

Mario knew exactly what Giovanni was contemplating on doing. The Calcagnis were thinking of the old woman and what the Goumiers did to her. There was no time for compassion and no need for religion at this point. These men were going to beg for mercy. Giovanni cut their trousers open, their genitals were completely exposed. They struggled but the old men were blessed with strength from a lifetime of work in the fields. They held the prisoners firmly as Giovanni repeatedly asked them how many people they raped.

They begged for mercy as Giovanni took an accurate swipe and severed the penis of the first man. The prisoner screamed as Giovanni repeated the procedure with the second Goumier. It took him two swipes to complete the job. The screaming rose as the men dropped them to the floor.

The Calcagnis Meet and Clear The Area of The Enemy

Angelina turned her head away. Giovanni said, "Let them live so they can suffer for the rest of their lives."

Mario looked at them and said, "Oh, what the heck." He opened fire and put them out of their misery. He ordered the old-timers, "Pull these bodies out of here and throw them deep into the ravine. The Allies will someday blame the Tedusci."

Angelina went back to the farmhouse and told the women what had transpired in the barn. The women shrugged with approval. Anna said confidently, "The shoe is now on the other foot, Mario is the equalizer."

"He can't stay here forever, he is in the military," Angelina replied.

The Calcagni's were sitting outdoors. There was no shooting in the area but they saw flashes of artillery from Ceprano and Arce. Luigi said, "It could be around the area of Frosinone."

Giovanni spoke with sadness, "A lot of people died and many more will perish and except for our mother, your grandmother we would have survived with only a few bad memories."

"The woman is saintly. She is a better person than the Pope and all his Bishops," Anita added.

"Mama has never missed a day of church unless she was sick. I never saw her without her rosary beads," Luigi said.

Mario offered some consolation, "Mama will recover. As far as this area is concerned, we have the edge. Between the trouble the Goumiers are facing with their officers and our element of surprise, we are in the driver's seat now."

Giovanni asked with some concern, "Will you be in any trouble leaving your unit?"

"Not really. I have a colonel-friend who is covering for me. In fact, by the time I get back he will in all probability be a general."

"Why will he cover for you?" asked Luigi.

"I saved his life and he is grateful. He sort of allowed me an official leave of absence."

"Saving the life of a colonel is a very big accomplishment in time of war."

"Yes, but I wouldn't save a person because of rank, I would give a private the same consideration."

Giovanni said, "Let me ask you a hypothetical question. If you were faced with helping a private that was badly wounded and a colonel who was wounded but not critical, who would you save?"

Pico

Mario thought awhile. "The colonel is more important to the war, but as far as determining who lives, they are equal. The private would get my attention because he is more in need of care."

Luigi assessed, "That is perhaps correct, but it won't get you any battle awards. Your uncle received the Iron Cross on the Russian front."

Mario shook Giovanni's hand and congratulated him. Giovanni quizzed Mario further, "Did you ever save the lives of any enlisted men?"

"Yes, I have. As a matter of fact, it was during the battle of the Rapido River. The Germans clobbered us. I was lucky enough to stumble onto a machine gun and kept the enemy busy until the men got back over the river."

"How many got back safely?"

"I believe it was close to a hundred. We sure were lucky."

"What happened to you?"

"I'm here, alive and well. I only caught some shrapnel."

The uncles were moved with his sense of modesty. Luigi asked, "What kind of an award did you receive for that effort?"

"The Distinguished Service Cross."

"I believe that's the second highest award given by the United States, isn't it?," Giovanni said.

"Yes, it is." He pulled the ribbon out of his jacket pocket and showed it to them. Another ribbon fell to the ground as Luigi picked it up and asked, "What is this one?"

"That is the Congressional Medal Of Honor."

Giovanni jumped up in surprise, "Jesus Christ! The most important award in the world."

"Uncles, my family is dearer to me than all the medals in the world."

"Thank God you feel that way. We owe our lives to you."

Mario grew uncomfortable with the flattery, and his uncles noticed.

Giovanni asked, "You got the Congressional Medal of Honor saving the Colonel?"

"Yes, along with a few others."

They were digesting all the exploits they could about Mario. After all, he was their nephew and they felt they had a right to know. Mario didn't want to talk about his accomplishments

196

The Calcagnis Meet and Clear The Area of The Enemy

anymore but respected his uncle's curiosity. Giovanni continued, "How many soldiers did you save?"

"It's difficult to determine because a great number of them would have become prisoners. The Germans were counterattacking in force through an important ravine. If they had been successful, they would have cut off about five thousand men. I had a squad and a radio with me and I knew the terrain well. They attacked after midnight and I caught them by surprise. I directed artillery fire on them and stopped them in their tracks. Very few escaped. The artillery followed the few that retreated to their original positions."

Giovanni told Luigi about the peasant he met near Pontecorvo who knew about the G.I. who was responsible for stopping nearly one thousand Germans. He was a mule skinner working for the Americans. Luigi confirmed, "I heard the story from someone else but nobody believed it."

"Were you that soldier they were talking about?" Giovanni asked.

Mario rose and stretched. He laughed and replied with a mock guilt, "Yes, I am that same guy." They looked at him in awe and acted as if they were in the presence of a god.

He spoke solemnly and seriously, "I'm not proud of what I did. There is no honor in killing. War is war. Those German men had families and I took their lives. They had no chance. I don't think I can ever erase the episode from my memory. The carnage I left behind in that ill-fated ravine will remain with me forever."

"When you go home, they will hold parades for you," Giovanni said with pride.

"Maybe and I'll try to avoid it. The president is going to award the medal to me. In fact, my commanding general doesn't know where I am. If he ever finds out, the Colonel is in big trouble. Please don't tell anyone here about our discussion. I have enough facing me back home. I don't want the same thing to happen here."

"We understand," Luigi and Giovanni concurred.

Mario further added, "I would love to stay here for a few weeks but too many people will get into trouble when the General starts asking about me."

"How about you, will you get into any trouble?"

"Maybe, but the citations will help. I want to leave sometime tomorrow so it is imperative that we take care of business tonight.

Pico

It might bother me to kill Germans, but I have no feeling when it concerns these Goumiers. My objective is to clear the area and drive the stragglers northward."

"We're with you," the uncles responded.

Mario turned to Luigi, "You're out, that leg will not permit it. You can be more helpful here at the farmhouse."

"He's right, you can hardly walk," Giovanni said.

"Have you guys ever heard of Pietro Colombo? He is a friend of mine."

"Sure we know him. He came here and retired with his family. Mr. Columbo has a farm somewhere between here and San Giovanni."

Mario asked, "I wonder if he's all right. I worked for him as a kid."

"He may be safe because he is old and had money. Pietro is a tough old bird."

"He is both but I got along very well with him. He was a good friend of your brother. I've got to check on him," Mario said with a smile.

"I've heard that there are quite a few Goumiers in that area," Giovanni said.

"Fine, we can take care of some business there. The more we kill, the better chance they will retreat back to their units."

"Let's go," Giovanni said.

The Calcagnis gave precautionary instructions to the people in the farmhouse and left together. Skirting the northern edge of town, they passed throught the cemetery. There was no evidence of any Goumiers in town. The French were there and some of the civilians were returning. However, problems continued to exist in the surrounding hills.

It took them nearly a half-hour to reach Colombo's farm. They were on a hill observing the property from about one hundred yards away. There was something afoot. It was nearly two o'clock in the morning and every once in awhile, a few stragglers entered the farmhouse. Mario peered through his binoculars. The house was lit and smoke emitted from the chimney. The door was missing and Mario distinguished women who were not Italian. He estimated there were twenty or more Goumiers in the building in addition to a half-dozen women who were dressed in the same brown colored

The Calcagnis Meet and Clear The Area of The Enemy

robes worn by the soldiers. Mario told his uncle, "It looks like they took over the house and are living in it. But I don't understand about the women. Who are they?" He handed the glasses to Giovanni, "Take a look and see what you make of it."

Giovanni readjusted the binoculars and scanned the area as well as the windows. "They seem to be having a party. I didn't know that the Moroccans drank wine."

"Why not?" asked Mario. "It's free, isn't it?"

"I thought alcohol was against their religion."

Mario snapped back, "Religion! These fuckers only worship the devil himself. Where in the hell is Pietro Colombo?" Mario slumped back, "I guess we'll have to wait things out. Pietro and his family may be prisoners. If we go in, they could be caught in the middle."

"Let's hold back and see what happens," replied Giovanni . "I heard that some of those Goumiers had wives with them."

Mario laughed, "The Mexicans brought their wives along every time they started a new revolution."

"Here is how I evaluate this situation. We can blow up that house but I would hate to destroy my friend's house. I have a feeling that Pietro and his family have either left or are dead. It looks like these bastards are using the property as a haven. I don't think their superiors are aware of this place and even if they do, I don't care."

Giovanni was scanning the property as his nephew was talking and stopped his sight near the rear of the house making out a figure that appeared to be the body of a man. He nudged Mario, "Take a look near the rear of the house."

Mario looked, "You're right. It's a dead body and there are a few more near the trees behind it."

"That answers your question about your friend Pietro."

Mario agreed, "Colombo never took any crap from anyone and it figures that he resisted these bastards. Let's wait a little longer and see what happens."

Giovanni relaxed and found the waiting to be an opportunity to counsel his nephew. "When you get home, there will be a celebration. People will decorate you and stand in line to shake your hand. Politicians will have a field day and reporters will drive you crazy. Your private life will be on hold. Fast-talkers might exploit you. People are always looking for new heroes. The victors

Pico

are extravagant in honoring heroes. They are certainly justified in doing so."

Mario replied, "I know, it comes with the territory. What do you suggest that I do?"

"Are you asking for my advice?"

"Of course, I am. You're my uncle."

"First of all, I advise that we pass up on what is confronting us in that farmhouse. You are too important and shouldn't take the risk. It is possible that the odds are not to our advantage."

"Okay, I respect that, but, this is the last pocket of Goumiers in the area and they have conveniently assembled into a cozy nest. We wipe out the nest and the civilians are home free. We will attack them at four in the morning when they will all be asleep. Hell, they feel so secure, they haven't even bothered to post a sentry."

"Mario, what about the women?"

"Fuck 'em. They should have stayed in Morocco. Look what they did to our women. I'm going to do them a big favor."

Giovanni shrugged that his nephew would make it through the war safely and continued, "You should put up with the parades and celebration for a few days as a point of respect and then disappear for a while."

"That's not easy. America is big, and news gets around fast. It is nearly impossible to hide and besides, I wouldn't want to live like a monk."

Giovanni interrupted, "I have an idea but you may not be ready for it. Perhaps I should mind my own business, but I'm your relative."

"No problem, tell me what your suggestion is."

"It may sound ridiculous, but I recommend that you come back to Pico and settle here. In two days you have made more friends that I have made during my lifetime. In years to come, Italy will become a paradise again."

Mario pondered, "I'm an American. I love sports, baseball, football, basketball and the rest of our sports. I enjoy the movies as well as our government. I love everything that is American. Don't get me wrong. If I were born and raised here, I would feel the same and never consider leaving Italy."

"Mario, you are going to make a lot of money and be financially established. Perhaps you can consider living part of the time here on a yearly basis."

The Calcagnis Meet and Clear The Area of The Enemy

"You know, uncle, that's a good thought. That idea appeals to me. I realize that many of the new friends that I met will be exploited. But, my mother and father are Americanized and won't relocate."

"You're correct but remember, they didn't win the Medal of Honor."

"Thanks, uncle, I'll give it some thought."

"One more thing, Mario. Here, we have the greatest sport of all games. We have soccer!"

Mario laughed, "With all due respect, that game is not American and will never catch on in the States. I'd rather watch or play hockey. I promise you, uncle, I'll be back, one way or another."

Night crept in and everyone in the Colombo farmhouse appeared to be asleep. Mario felt that he was doing the right thing. All he could think of was the atrocities that were committed and the future mental and physical consequences. He had no difficulty associating the death of Italians from the horrific act he was about to undertake. The Goumiers in the farmhouse enjoyed the spoils of war. Mario intended that their days of callous exploitation would end before the break of dawn. He asked his uncle to take a position to cover the rear of the house.

Mario explained his plan. "They are all asleep and they don't have a sentry posted, which is stupid. We have a distinct advantage and they don't have a chance. I will give you five minutes to get into position. Don't shoot until I begin first."

"How are you going to handle it?"

"No problem, I am going to make my way up to the front of the house and throw in at least six grenades. After they go off, I'll rush in and finish them off."

"Why can't I contribute a little more than just watching the rear of the house?" Giovanni asked.

"If the word ever got out to their superiors that you participated, they could hold a firing squad for you. Conversely, if they ever find out that I was responsible, they may give me a medal. Besides, you are covering me. Who knows, the possibility is that someone could get out a window or back door and get behind me."

Giovanni said, "Now that you put it in that perspective, I agree." They embraced and he left for his new position.

Mario made his way quietly down the hill. He reached the edge of the clearing, a good one hundred feet from the doorless opening.

Pico

He guessed they probably used the wooden door to make a fire for cooking.

The next position was a stone well located nearly half-ways between him and the house. Mario could not see his uncle but assumed that he was in position. He crouched and ran as he reached the well and took a moment to check his grenades and weapon. Mario was about to proceed when he noticed a foot partly visible in front of the small stone structure. He crawled around the other side and discovered a body lying face down.

Mario turned it over and noticed that it was his old friend, Pietro. His throat was slashed. Mario made the sign of the cross and bit his knuckles in anger.

He moved stealthily to the house and reached the side of the doorway. The outer walls were thick and sturdy and consequently would offer protection from the blast of his grenades.

The interior included a large room which had a huge fireplace. Chances were good that the occupants were sleeping on the floor near the fireplace. He could hear snoring and an occasional grunt and stirring. For a moment, he doubted if he could destroy the massive structure.

The element of surprise assured him that there would be no survivors from the grenades.

Mario laid out six grenades in front of him and placed his weapon within reach. He said to himself, "It's now or never."

In rapid succession, he removed the pins on four of the grenades, waited a moment and rolled them into the large room. As the grenades came to a halt, he removed the pins from the two remaining ones. The timing was perfect. The four grenades exploded almost in unison and blew out the interior of the house. Mario threw the other two grenades and they went off immediately. He sprang into the opening and fired his submachine gun at anything that moved.

The explosive power of the grenades was deafening.

The accompanying smoke was thick as Mario made his way carefully to each room. Everything happened so fast that the inhabitants of the rooms were either dead, dazed or unable to react.

Mario tossed more grenades in each room. He heard shots for the first time from outdoors.

Mario was backing out when he suddenly felt a sharp pang in his left thigh. The shot came from behind him. He whirled and fired at the same time. One of the occupants had survived. He clutched his weapon as Mario finished him off.

The shooting outdoors ceased as Giovanni had made his way down to the rear of the house. As he neared the front doorway he called out, "Mario!"

"It's okay, I'm coming out." He came out limping asking, "Did we get them all?"

"Yes, nephew, I shot three of them in the rear of the house. You've been hit!"

Mario sat down. "I guess I'm losing my touch. I can't see how anyone could live through that barrage!" He took off his white scarf and Giovanni wrapped it tightly around the wound.

"If it's any consolation, you are fortunate, the bullet went completely through without hitting any bones. Kid, you're as lucky as your Uncle Luigi."

"Let's get the fuck out of here. The French soldiers in town will think that the Germans are counterattacking."

Giovanni said in jest, "You did the counterattacking. How many do you think are dead in there?"

"I don't give a fuck. By the way, Pietro is dead near the well. They slashed his throat."

Giovanni added, "The civilians in the back also had their throats cut. They are the remains of your friend's family. The women were raped before they died. The front of their clothing were torn to shreds."

Mario rose with help from Giovanni. They moved back into the safety of the hills.

Mario limped through out the trek. The bleeding stopped, but the pain intensified. He knew that some rest along with medical treatment from his first-aid kit would help. They stopped and rested and they could see the Colombo farmhouse from a distance. The house was ablaze. Mario remarked, "We made the right decision, Pietro and his family would not be around to use the house."

The Calcagnis were making their way back as Giovanni informed Mario, "We can cut through this crossroad area and save some time. There's a path on the other side that leads to the rear of the farmhouse."

"Okay, let's chance it."

They looked around and were convinced that the area was clear. Giovanni assisted his nephew as they hurried across but were halted

Pico

by a grotesque scene. Ten bodies were stacked in a heap. The bodies were dismembered. The torsos were in a separate pile and the heads, legs and arms were strewn haphazardly along the intersecting roads. The deteriorating bodies emitted a ghastly odor. Upon closer observation, they determined that the victims were Goumiers. Mario asked, "I wonder what happened here?"

"I'll tell you what I think as we get the hell out of here."

Mario agreed.

Giovanni said, "The Germans did it, probably during a counter-attack. They must have caught them raping prisoners."

Mario replied, "It makes sense, I found a similar situation regarding American soldiers."

"No kidding, what did you do?"

"They tried sodomy with American soldiers and now reside with their Allah."

Giovanni asked, "What's wrong with these people?"

"I don't know, but from what I hear, most North Africans are normal. It's only this group, known as Goumiers who were trained by an insane general who has allowed them to do as they please. For them, war justifies their primary motivation which is rape, pillage and plunder."

"They are a disgrace to humanity. They received their just due by being hacked and dismembered with their own long knives," Giovanni answered.

"I agree with you one-hundred percent. They were on the receiving end of justified vengeance." The Calcagni team arrived at the Micheletti farmhouse as daylight was breaking. Two men were awake and standing guard. Mario smiled, "Good work men, you are real soldiers." He shook their hands.

Anita patiently waited for her husband.

She worried constantly for his safety. Luigi had told her repeatedly not to worry. Giovanni proudly related Mario's heroic exploits to all.

Anita screamed, "Mario is hurt!" Anna leaped to her feet. Angelina was soon helping Giovanni with aiding Mario.

"I'm fine, a little rest and some medicine is all I require."

The women set him down gently and administered to his wound. They bandaged his leg and he fell asleep. Giovanni recounted the entire battle to the occupants who were now all awake. Angelina beamed proudly. Luigi said, "The boy is a one-man army."

The Calcagnis Meet and Clear The Area of The Enemy

"I was there. If I didn't see it for myself, I would never believe it. He demolished the area," Giovanni agreed.

The women agreed. "Thank God for Mario. He is a saint."

Luigi laughed, "I wouldn't go that far, but here he is a bigger hero than what he did for his own country."

"It's sad because whatever he accomplished for us must be kept quiet forever," Giovanni rationalized.

Gabriella was sitting next to her sister. "No one can ever hurt him."

"Mario will be bigger-than-life in America," Luigi agreed.

"Yes, that's true," replied Giovanni, "but let's all keep it quiet."

Anita added, "Yes, like a Sicilian."

Angelina watched as Mario rested. Her head was propped in her hands as she day dreamed that they would someday be together forever.

22

THE HERO DEPARTS

Mario had planned to leave early that morning but the Calcagnis purposely did not awaken him. His uncles decided to let him rest as Angelina spent the day sitting next to him.

Anna told Gabriella with some concern, "Your sister is going to receive a rude awakening. Mario will never be back, the girls in America will see to it."

"Mama, you don't know Angelina and Mario doesn't know her like I know her. Mario is in love with her but it hasn't blossomed yet. He doesn't realize it yet and won't understand his restlessness. Whenever he meets an American girl, there will be something blocking the way. He will never get serious."

Anna was deeply moved by Gabriella's assessment and asked, "How do you know these things?"

"Mama, like most people of this area, I believe in dreams. I once dreamt they were together after the war. They were walking in the fields with three small children running playfully. Mario and Angelina were singing 'Santa Lucia.' Mama, he is the best singer that I have ever heard."

Anna wanted to believe Gabriella's dream but doubted its possibility. There was no way that her frail-looking Angelina could compete with the beautiful American girls. With his reputation, he certainly could have the best.

Angelina, however, was more intelligent that any young girl she ever knew and a very determined individual who's only handicap was being thirteen years of age.

The Hero Departs

Gabriella broke her mother's train of thought, "Mama, I know that you believe in dreams."

"I do and would love to see it all come true but you are conveying the impossible." She looked at Angelina, still on vigil, and then at Mario asleep and thought to herself: My poor little Angelina, she is going to end up in the crazy house.

It was mid-afternoon, the sun was shining brightly as the villagers began to feel that peace could very well have arrived. The road ahead was difficult and the wounds of war would heal with time. The immediate problems facing them, however were lack of water, food and disease.

Mario slept the most of the day and awakened at four in the afternoon. He was surprised that he slept so long but it was the first real rest he had in days.

As soon as he stirred, Angelina arose and went outdoors. Most of the people left and had gone back to their own homes. Giovanni assured them that it was safe. Mario tried to rise but his leg gave way and he nearly fell. Anna ran to him and he braced himself against her.

She walked him outdoors where they were met by his uncles. The first thing he asked was about his grandmother. They had moved her from the attic and had her seated under a fruit tree. Luigi limped over to him and said, "Go see her yourself, she is doing fine."

Mario smiled, "We each have one good leg, perhaps we can hang around together and assist each other. Where is everybody?"

"Most of them headed back to their homes to see what is left," replied Giovanni. "You had a good day's sleep."

"I know but I slept too long, I've got to get back to my unit."

They recommended waiting until the following morning insisting that he rest for an extra day. Anna said, "Why travel in the dark, when it's safer in the daylight."

Mario reached out for his grandmother and kissed her on the cheek. She smiled, "My Rocco, I knew that someday you would come back."

"Yes, Mama, I will be here for a while but when I leave, I promise you that I will return." The sons were satisfied to see her recovering. She smiled most of the time and appeared happy to be alive.

Pico

Darkness came and Mario agreed to spend the night but he declared, "I must leave in the morning."

They all sat outdoors and had a light dinner at a large table under a grapevine.

Earlier in the day, the women foraged for wild broccoli and dandelions. They made a fire for the first time in many months and cooked the vegetables to make minestra.

Plenty of wine was evident and the meal was appreciated by all.

The conversation was pleasant as Mario told them about life in America. Some of the children thought that in America money grew on trees. Mario laughed and rubbed the toddler's head with his hand. He told them that fruit trees produced money.

Angelina was very quiet. She starred at him in the darkness. Her hero would be gone in the morning and there was nothing she could do about it. Gabriella saw the sadness but kept her distance.

Giovanni asked Mario, "With that leg, how are you going to get back to your unit?"

The American gave a deep sigh, "I've been giving it some thought. I know I can't walk back. Once I get on the road to Itri, I should be able to hitch a ride. By now, there should be some traffic along that road. I've also thought of going into town and asking the French to give me a ride back to Naples."

Luigi asked them both with concern, "Isn't that dangerous after killing a lot of their Moroccans?"

"How the fuck would they know? We all know that the Germans did it."

Giovanni agreed, "Nobody has ever seen Mario. They don't even know he exists."

"I think my best bet is to get down to the road."

Anita offered an excellent suggestion, "We should be able to find an asino somewhere and Mario can ride it until he finds a better way to get to Naples." The children were ordered to look for a stray mule or jackass.

Giovanni inquired, "I thought you were going north to join your group."

"It's too late for that, they must be in Anzio by now. Besides, I didn't plan on getting shot in the leg."

The night was quiet. The nightingales were singing as the peasants looked serenely into the clear night.

The Hero Departs

The Allies had sprinted northward and were preparing an offense to liberate Rome. The French Divisions were in the mountains north of Pastena and they were not stopped. The plan was to keep them out of the liberation of the Eternal City. Mario realized the reason for the Pico problem. If the high command had allowed the French to advance, perhaps some of the atrocities could have been avoided.

The children came back with a stray mule. Mario would now be able to travel better out of Pico.

They were talking and enjoying themselves when Anita interrupted them, "Quiet, I hear some rumbling in the distance."

Giovanni replied, "It's coming from Pico. It sounds like truck movement. Perhaps the French are moving out."

Giovanni and his wife climbed a nearby hill to get a better view and noticed a truck convoy moving in from the Itri road. They ran down and told Mario.

As everyone was contemplating who they were, a youngster came running up the lane shouting, "The Americans are here, the French are leaving and the Americans are taking over Pico. Everybody is celebrating."

"Well, I'll be God-damned. Now I'll have to change my plan again. All I need to do is go into town," Mario said.

"Too late, Mario, it's one in the morning, get a good night's sleep," recommended Giovanni, "Leave tomorrow." Mario nodded and they all retired for the night.

Mario was having difficulty sleeping. There was something in his mind but he couldn't bring it forth. He thought about the serenity of the foothills and knew that within a few weeks every event he would face would be in direct contrast to this quiet and pleasant night. Everything would soon move fast. He slept in brief spurts. There was something in his mind he could not nail down and remained elusive.

Dawn broke and the sun was rising over the Abruzzi mountains as Mario awakened to the sound of voices outdoors. During the night, everybody who knew Mario was told that he was departing in the early morning. The peasants were assembled and waiting for him to appear.

Giovanni helped him to his feet. "It's your first big day, you may as well get used to it." The American washed his face in a

Pico

basin and donned his gear. Giovanni carried his weapon for him. Mario stepped outdoors to a cheering crowd. The women had plucked their favorite yellow mountain flowers that they gave him as he limped through them.

He shook every hand as the men embraced him and the women kissed him. Angelina, Gabriella and Anna were at the end of the line with his family. He stopped in front of Anna who removed a gold necklace from her dress pocket and placed it around his neck.

He said, "Anna, that's not necessary, you must take it back. The gold will buy you food." She said, "Thank you for our lives and think of us every time you wear it and don't worry about us, God will provide." She embraced him and kissed him on the cheek, "We shall love you forever!"

Gabriella hugged him and handed him flowers along with a loving smile. He knew that the beautiful young lady had recovered from her ordeals. Angelina looked at him with tears. She awkwardly offered him flowers and he pinched her cheek, "Take care of yourself, little lady," he said and embraced her.

Angelina tenderly stepped back shyly with her head down.

Anita and the children swarmed around him. He promised to write them and that he would return.

Luigi embraced him tightly, "Give my best to my brother, I was only six years old when he left." Luigi cried unabashedly and sadly.

Mario knelt before his grandmother as she ran her hand through his hair and smiled, "Goodbye, my Mario, and be sure to visit again soon."

"I will, Mama, I promise you, I will."

Mario limped a few feet towards the mule when Anna called out, "Mario, can you sing?"

"Yes, I think I'm good enough for the opera."

She smiled broadly as the American approached the animal when Angelina shrilled out "Mario!" He turned as she leaped into his arms and embraced him with all her strength. She placed her cheek against his and wouldn't let him go. He patted her back as he felt her tears trickling between their cheeks. Mario did his best to comfort her and said, "You are the bravest and most intelligent and the most beautiful girl that I have ever met." She turned her face to him and he gave her a little kiss on the lips and set her down.

The Hero Departs

Giovanni had decided to accompany his nephew into Pico. Without speaking further, Giovanni helped him mount the mule and handed him his weapon. He waved goodbye with tears in his eyes as Giovanni led the mule down the trail towards Pico. As they were half way down, they could hear the sweet, clear voice of Angelina calling out "Mario!" She ran from one ridge to another until her voice was out of range. Angelina finally dropped to her knees and cried hysterically.

The Calcagnis neared the bottom of the mountain and Giovanni said, "Whoever is strong and fortunate enough to marry that young lady will be the happiest man in the world."

Mario replied, "I don't doubt it, but she is still a child."

"She is a highly motivated woman in a child's body but the body will someday ripen to womanhood."

"Uncle, I agree with you all the way. When that happens, some Romeo will sweep her off her feet."

"That will never happen to Angelina, she is special."

They entered Pico and American forces were occupying the town. Soldiers were surprised to see an American infantry sergeant sitting on a mule. He rode into the center of town and asked directions to the headquarters. He made his way to the designated building and asked Giovanni to wait as he entered.

Mario approached a seated captain and identified himself. The officer rose with a smile and came around his makeshift desk.

There was a gleam in his eyes, "The Colonel, I mean, General Biggs told us to keep an eye for you. They just promoted him to General and he expressed concern about you."

"Boy, will he be happy to know that you're safe." The Captain saw the wound and immediately called for medical assistance. As ordered the wounded hero to be seated Mario interrupted, "I need a little favor."

The officer said, "Whatever you need, just ask."

"My uncle is outside and the rest of my family is located up in the hills. I want to be sure that they receive some food."

The captain instructed a sergeant to load a jeep with blankets and provisions and drive Giovanni back to his family.

Mario went outside as the medics arrived. Giovanni abandoned the mule as Mario approached him.

He explained about the provisions and the assurance that the Americans will give them special attention. They looked at each

Pico

other, shook hands and embraced for the final time. Giovanni instructed, "Give my best to brother Rocco. I was ten years old when your grandfather died. He was only sixteen and as you know, he left for America when he was seventeen years old. My brother, God bless him, always sent us money to help us. He was only a boy when he shouldered the responsibilities of the man of the house. Tell him for me that he accomplished his duties."

"I will Uncle Giovanni, and I will do my utmost to urge him to visit here."

Giovanni waved as his heroic nephew departed.

23

TYING LOOSE ENDS

The captain called General Biggs who was elated to know that his sergeant was well. He ordered that Mario be escorted immediately to headquarters in Naples. Meanwhile, the medics took him to the aid station and dressed his wound.

The people were celebrating as Mario was driven through town. Again, he was showered with flowers but had mixed emotions. It saddened him to leave Pico but happy to know that his relatives and friends were all safe. The Americans have arrived. The war was over for the Italians. Mario and his escorts drove towards Itri. The tanks that had blocked the road were bulldozed over the side.

By noon, they reached Cassino. The city was nearly leveled. Mario looked up to Monte Cassino and saw a few projecting walls. He told his companions, "On that mountain top, lies a sad relic of civilization. There was no need to destroy an irreplaceable structure of art." The driver nodded as they were halted by oncoming traffic.

Mario and his small group reached headquarters at two in the afternoon. General Biggs anxiously awaited his arrival and was out front when Mario got out of the jeep. The general placed his arm around Mario and said, "It's almost four days. I have been lying to the Commanding General all this time."

"What did you tell him?" Mario asked.

"I told him you were on R&R in Salerno getting laid every night."

"How did he handle that?"

"The son of a bitch shook his head and roared with laughter." The General and Mario laughed at the commanding general's reaction.

Pico

"Seriously, Mario, how did it go?"

"Not bad at all, sir, there were a few small problems but my family and friends are fine. I want to thank you for sticking your neck out. I now realize that you took one hell of a chance in my behalf."

"Mario, my son, if it wasn't for you, I wouldn't be here. I still owe you and I know that I can never even things out."

"Forget it, General, we are even."

He waved a hand and instructed Mario to relax and be seated. "I notice that you've been wounded. How did it happen?"

"Them fucking Germans kept counterattacking and I had a little run-in with them."

The general responded, "Now that you mentioned it, the French in the Pico area confirmed that the Moroccans had a tough time with counterattacks. Speaking of the Moroccans, how did you make out with them?"

Mario shifted in his chair and looked the general in the eye. "I took care of business and I don't want to talk about it." The general nodded and shuffled some papers.

"I'm happy about it but why are our men in Pico?" Mario asked.

"We've decided to cut the French Mountain Divisions loose providing they advance in the rough mountain terrain only. They belong where there are no humans around. Since your absence, another event regarding you is afoot."

Mario looked up, "Now what?" An aide entered with lunch and coffee. The general thanked him and he left.

"This is the first good meal I've had in nearly a week," Mario said.

The general sipped some coffee and sat back. "It appears that some politicians back in stateside area are causing a ruckus over your Distinguished Service Cross."

"What's their problem?"

"They believe that you were shortchanged and should have two Congressional Medals of Honor. Some attorney general and a retired navy commander are getting a lot of important people on their side."

Mario appeared to be irritated, "I guess they don't have enough to do. You know, I'm not looking forward to going home. It's going to be a three-ring circus and I will be on display forever."

Tying Loose Ends

The general placed his index finger to his mouth and explained, "For whatever it's worth, I also believe that you were short-changed."

Mario threw his hands up in the air and expressed himself exasperatedly, "I'm getting sick of all these medals and I haven't even officially received them yet."

"Calm down, son, a month from now everything will be back to normal for you."

"No, General, you're wrong, it will last a lifetime." He didn't reply but poured Mario another drink. He calmed down and said, "You know, I can't blame them for wanting to upgrade the award. From their prospective, they do have a reason."

"What do you mean by that?"

"Those two individuals, I love them, I owe them and believe me, I appreciate their efforts."

"Mario, you don't owe anybody anything." The sergeant slumped back in his chair and bit on a stalk of celery. He was contemplating whether or not he should continue the conversation. Mario admired the general but felt that his personal life was his business.

"Son, you can tell them all to go to hell."

"General, before entering the service, I was in a little trouble. The Armed Forces didn't want me. I was a fucking felon. My life was at a crossroad and the only chance I had to straighten out was to go into the service. We had a retired naval commander in our area who knew me since I was a child. Sir, I was fortunate that this man took an interest in me. I didn't realize it but he had a lot of clout. We were all aware that he was awarded the Navy Cross during World War One. I had no idea but his brother-in-law was the Attorney General of the state. Let me tell you, those Irishmen have a tremendous amount of power in New England."

Mario paused as the General filled both their glasses with cognac and said, "Go on, son."

Mario continued, "The Commander took it upon himself to organize a committee and obtained signatures in my behalf to form a petition. We both met with the Attorney General and submitted the petition which had about five hundred names on it."

"Five hundred signatures is a lot of people. You sell yourself short. That number of people would certainly indicate that you have a lot of friends," Gen. Biggs replied.

Pico

"Now that you bring that up, that is five hundred more people who would appreciate seeing the upgrading of my first award. Well, to make a long story short, the Attorney General wiped out my record and cleared me for immediate duty. They expected big things from me and I promised them what they wanted. I never had any pressure thrust on me to perform. The only thing that I ever joined in my life was the Boy Scouts and I have been a strong advocate of their motto."

The general said, "I was an Eagle Scout." Mario replied nonchalantly, "So was I. I have been fortunate to be in the right place at the right time. It happened to me on two occasions. The first time, I happened to be close to a machine gun and did what most soldiers would do under the same circumstances. I fired the weapon and threw a few grenades and then ran for my life. The second time, you wouldn't believe it, but I took my squad along the wrong trail and you happened to come along. If I didn't have a radio, neither one of us would be here. So you see General, I had a radio and was lucky to be in the right place. It's all a matter of timing. The other thing I have going for me is that my mother prays for me all day long and of course, I've already explained about my dream."

"Mario, I can't figure you out and I'm not going to try. You have more guts than anyone that I've ever met. You are dedicated and absolutely fearless."

"General, you lost me, I don't know what the hell you're talking about."

"Remember one important thing. You are an American hero. What you have accomplished overrides anything from the past and present."

"Sir, I only want to be happy, healthy and live a quiet life."

"Leading a quiet life won't be easy, but do the best you can." The telephone rang and the general excused himself and answered. Mario tried not to eavesdrop, but caught the word Pico along with Goumiers and the French.

The general was talking to some colonel who was in charge of the Pico area. They conversed a few minutes as General Biggs thanked him for the information. He hung up the telephone and started laughing. Every time he tried to talk, he choked on his words and his face turned red.

Tying Loose Ends

Mario poured him a glass of water and said, "What the hell is so funny?"

General Biggs regained his composure and asked, "What in the hell did you do out there in Pico?"

"Why do you ask?"

"The French found over forty of their fucking Goumiers dead and they say that the Germans didn't do it."

Mario replied in defense, "I heard that the French themselves are shooting Goumier renegades that they can't control."

The general began laughing again and said, "Mario, you are, without a doubt, a warrior for the ages. The French division investigators found American submachine gun shells all around a destroyed farmhouse. They also found American-made grenade fragments."

Mario laughed along, "When I left Pico this morning, there must have been at least four hundred Americans in the area and they all had guns."

Gen. Biggs said, "That's funny. One Goumier escaped from a nearby mountain cave and testified that the American was wearing a white scarf."

"Those bastards killed at least a dozen people that I knew or knew about. They beat and raped too many to mention. The thing that hurt me the most was that they raped my eighty-year-old grandmother six times," Mario countered.

Mario had tears in his eyes as the general came around and comforted him. "I'm on your side, but between you and I, did you take care of that business?"

"You can bet your ass that I did."

"I congratulate you. If I were in your shoes and had your ability, I would do the same thing."

"General Biggs, the outcome against those Moroccans was never in doubt." The senior officer said cautiously, "Well, its over now, I'm not letting you out of my sight. Your next stop of address is going to be the United States of America."

Mario laughed, "Hey, General, why don't you come along?"

"I would love to, in fact, I'll give it a try."

The sergeant replied, "Before I go anywhere, I need one more special favor," the sergeant replied.

"Son, you can have just about anything you want."

Pico

"I need to find the German prisoner who saved my ass on the road between Itri and Pico."

"How do you know that he is a prisoner?"

"Because I talked him into surrendering and directed him to the American lines near Itri."

"This is highly irregular but what is his name?"

"Horst Schultz."

General Biggs picked up the telephone and fifteen minutes later he was informed that he would receive the information he requested.

"How did he save your life?" the general asked.

"He shot my horse instead of me and held me prisoner until he noticed my white scarf with my name on it."

The General wasn't concerned about the horse. "Why did he let you go?"

"Because he knew that I was going to Pico and he has a girlfriend there that he wants to marry."

"Tell me if I'm wrong. You want us to release this prisoner so he can go back to his girlfriend."

"Yes, that's if he wants to."

The general shook his head as the telephone rang. He was told that they did have a Horst Schultz as a prisoner. The General ordered them to bring the German to his office. He was told that it would take about thirty minutes.

Mario heaved a sigh of relief and said, "I thought that he ran into some renegade Goumiers."

The general shook his head, "You've got more lives than a cat. The German could have shot you instead of the horse. He had you prisoner and I'm under the impression that you were on a cake walk."

"I guess that we are both lucky, besides, the German troop was weary and a few of them were wounded. I had a very good chance of jumping them. General Biggs, I am sincere when I ask you to come with me. They didn't give you an infantry division and this headquarters duty is a bunch of crap. Your battle record speaks for itself, they promoted you to general. If you get a chance to meet the President there is a good chance for another star."

Biggs thought awhile and said, "Mario, you've got a lot on the ball.

Tying Loose Ends

I can assure you right now that the other award will be upgraded and no one has ever won two Congressional Medal of Honor. Yes, Sergeant Mario Calcagni, I'll be happy to ride your coat tails to another star. All I need to do is convince the Commanding General."

There was a knock on the door and a sergeant stuck his head in and told them that the German was present. He was ordered to show the prisoner in.

Horst entered and saluted the General. Mario rose and welcomed him by shaking hands. The German beamed and told Mario, "Its good to see you alive."

"Thanks, I am happy to see that you took my advice. How are they treating you?"

"Very well. Did you see Gabriella?"

"Yes, Horst, she is fine now."

He expressed concern and asked, "Why, did something happen?"

"She had a tough time like everybody else but she is doing well now"

"Tell me, what happened?," asked Horst.

Mario placed his arm around the German's shoulder. The General knew that he wasn't just looking at an American and a German. He was seeing two close friends. They were touched by the war and survived. Mario spoke with hesitation, "Gabriella was raped by the Goumiers."

Horst began to weep and explained, "I wanted to leave the Army and take that wonderful family into the mountains but the mother refused."

Mario consoled him, "Gabriella loves you very much but she thinks that you no longer want her."

"Those Italians are so sensitive. I shall love that girl forever. What happened to the bastards that raped her?"

"I took care of that situation," Mario replied.

The general told Horst to be seated and gave him a glass of cognac. He accepted and said, "Gabriella and I have been through a lot. How is Anna and Angelina? How is Mama, I mean, your grandmother?"

"Anna is strong, she was mauled and raped. My grandmother suffered the same fate."

"They did that to Mama?"

Pico

"They did that to most of the people in the entire area, but it's over now. The victims must forget and forge ahead. That is where you come in. I want you to go there and help these people."

"I promise you Mario, as soon as this war is over and I'm released, I'm not going back to Germany. I'm going to Pico and marry my Gabriella. By the way, how is little Angelina?"

"Don't worry about Angelina, she killed two Goumiers with the shotgun you gave her."

He placed his hands to his head, "Angelina is Angelina. She is fearless and indestructible. The only man who will be able to tame that young lady must be stronger than her. Did you know that she faced down the German Commandant in Pico and saved the lives of more than thirty civilians from the firing squad?"

Mario smiled, "I never heard that story."

"Angelina will become a legend."

Mario looked at the General and said, "I guess that you are aware of a favor I'm asking."

"Yes, I know what you want, but it's impossible."

"Nothing is impossible and you said that the world is at my feet. I would like Schultz transported to Pico with the proper papers. His war is over and I owe my life to this man. In fact, you also owe him a favor."

The general got up and paced the floor. He querried, "How do we account for him?"

"I don't see any problem, the man speaks near-perfect Italian. He wanted to defect and live in the mountains. This man has a big heart, General Biggs, if you want to do me a favor, cut his orders and let him go to Pico."

Horst knew that Mario was important but couldn't understand how he could order a General around. He kept quiet and kept his fingers crossed. The general walked out the door and gave some orders to his first-sergeant. He returned after a short time and told Mario that his wishes would be granted. Horst would be on his way by morning. Mario shook his hand and thanked him.

Mario faced his German friend, "Someday I will return to Italy and we will share a bottle of wine together. Be friends with my uncles and take good care of the Micheletti family." The two former adversaries embraced.

Tying Loose Ends

The German saluted and thanked the General. He was led away for processing and ultimate release. General Biggs smiled and said, "We did the right thing. Now I will get the ball rolling to get you back to the States."

"You mean, get us back there, my General."

24

STATESIDE WELCOMES THE HERO

Everything was falling nicely into place. Mario's Distinguished Service Cross was added to a second Medal of Honor award. Political power and public opinion were helpful in that it forced a review.

The determining factor was that Mario's division commander took a great deal of flak because the division was soundly defeated. Some say that he was the fall guy, but the general wasn't in the right mood. It was an oversight and he willingly agreed to recommend Mario for the Medal of Honor.

The word spread around Stateside that the great hero was finally coming home.

It was decided that General Biggs would accompany Mario for the ceremonies. The Army top brass thought the move would be ideal for Army public relations. The plan was to leave within a few days for Washington, D.C. The president was looking forward to presenting the awards to Mario.

His family, along with some close relatives and friends, were invited at government expense, to attend the ceremonies. In addition, Patrick Morgan and the Attorney General along with their wives would be in attendance.

On the morning of departure, the division band performed for the occasion. The commanding general made a brief speech. He thanked Mario and stated that he had but one regret. He would have been honored to present the awards to him but he had to defer to the Commander-in-Chief. The general explained that every recipient of the Medal of Honor deserved a place in history. It was his opinion, however, that the valor of Sergeant Calcagni on the Italian mountain could very well have been the turning point in the course of the battle. The enemies had clear advantage but were rebuffed. They never regained momentum. He saluted Mario and shook his hand. The band played as the sergeant boarded the aircraft.

General Biggs sat with his aides and noticed that Mario had tears in his eyes. The plane taxied to the runway and made a final check before takeoff. Mario strapped on his seat belt as the engines revved up . He was offered a magazine with a marine war hero on its cover. The plane powered down the runway and lifted off into the sky. Mario was pleased knowing he was spared from the long and rigorous home-coming by sea. They made a sweeping left turn as the aircraft continued to rise. He looked below and caught sight of the destroyed abbey on Monte Cassino. It was easy to follow a line along the Aurunci Mountains as he looked to the right and observed Pico. Mario thought: It appears to be peaceful, as if nothing ever happened there. Mario kept his eyes on Italy until it was no longer in sight and eventually fell asleep.

The flight would take them over North Africa, from there to England. The third leg was over the ocean to Newfoundland and finally, Washington D.C. After a number of restful hours, Mario awakened at sunset. He asked the General, "What happened to North Africa?"

"We bypassed it and are heading for England." He slapped Mario on the knee and asked, "How are you doing, son?"

"I'm fine. How long do we stay in England?"

"I believe we only take on fuel and then we're off to America. On arrival in New York we will get a chance to shower, shave, catch a good meal before taking off again for the Capitol."

London was heavy with fog and they were rerouted to a military base near Scotland. After landing, the entourage deplaned and stretched their legs. Coffee and doughnuts were served as the aircraft refueled.

Within the hour, they were airborne. The turbulent updraft periods caused a rocky and rough trip. There were times when Mario thought that the aircraft would go down. It was his first flight so he didn't have the experience of his fellow passengers. The first leg of the trip was smoother but flying over the north

Pico

Atlantic was another matter. Biggs noticed that Mario was uncomfortable and observed him griping the seat with such force that his knuckles turned white. He smiled at the sergeant and said, "There is nothing to worry about, I have been in worse conditions. Hell, this isn't even a storm." To get Mario's mind off the rough flight conditions he continued, "When we arrive in Washington everything will move fast for a few days."

"Tell me about it," Mario asked.

"We will be traveling by motorcade directly to the White House where the President will present the two awards. There will be a ceremony then a state dinner in your honor by the President."

"What about my family and friends?"

"Yes, Mario, to my knowledge, more than twenty of your people have been invited."

The plane suddenly dropped altitude by fifty feet and Mario tightened his grip on the seat. "It should be a great experience for them to meet the President and dignitaries."

"I hope they don't make fools of themselves."

"Don't despair, they will adjust to the situation and the public relations people will guide them and keep them at ease."

"What happens after Washington?"

"We are going to your hometown and I understand that a parade is scheduled for the occasion."

"Then what?" Mario asked.

"After you settle down with your family, I'm leaving for a few weeks. It will be vacation time and then we meet in Boston. We will be with a group of dignitaries on a war bond drive. You, my friend, will be the main attraction."

"Fine, if I make it to that point, the rest is easy. Who are the people that will accompany us?"

"A few celebrities, some United States Senators and whoever joins us during the tour."

"Who is in charge of this operation?" Mario asked.

"Who else but yours truly."

One of the General's aides brought them box lunches along with a thermos of coffee. They had traveled over half-way across as Mario debated whether or not he should try to eat and decided to pass and drink only the coffee. The general offered, "It's roast beef on rye, the British are famous for their roast beef."

Stateside Welcomes The Hero

Mario replied seriously, "I don't want to use that bag to throw-up in."

Gen. Biggs was enjoying his meal and said, "If you think about it, that's exactly what will happen."

"I'll save my dining for our layover in Newfoundland. How long before we arrive?"

"The estimated time of arrival is about two hours from now. We are carrying a full load."

"What are we transporting?" Mario asked.

"The hold of the aircraft contains the bodies of heroic Americans who are going home for burial."

Mario was silent. There was no need for words. Both men had seen enough casualties of war: dead or maimed.

This was, however, different. There was no war up here in the sky of the North Atlantic. An aide brought foward a bottle of Chivas Regal Scotch.

The General was elated and thanked the aide, "Where in the hell did you get this?"

"We have friends in England and they knew about Mario. They gave us a full case."

The aide had two glasses for them and opened the bottle. He said, "Here, allow me to do the honors." He poured the Scotch over cracked ice.

The general said, "This is the best Scotch in the world." They consumed almost half the bottle as the plane touched down.

The weather was cool for late June but otherwise pleasant. After being hosted to V.I.P. treatment on the base, they retired for the evening. The next morning, the small entourage of military men were up early. The itinerary was altered as they bypassed New York. They had breakfast and departed for Washington in a chartered airline. Everyone aboard was in a jovial mood except for Mario. Being the main attraction in the upcoming events was wrecking his nerves. A beautiful brunette hostess served Mario and the General with cocktails. She gave the hero the warmest smile. He returned the gesture as she offered, "Is there anything I can do to make you comfortable?"

"Yeah, change places with me, I'm afraid of what awaits me ahead."

Pico

She laughed while running a hand through her hair, "I can't believe it, Sergeant Mario Calcagni scared like a little boy? I would give my soul to be in your shoes."

"Perhaps you can accompany me to ease the tension?"

She replied, matter of factly, "I'll be lost in the shuffle as soon as the ceremonies start." The hostess moved behind his seat and began slowly massaging Mario's neck and shoulders.

He relaxed and said, "Now that feels great. Do you do this for all your passengers?"

"Don't be absurd, only for men who have been awarded the Medal of Honor."

"How many have you met so far?"

"You are the first and I must say, a handsome hero."

She continued the tender loving care as Mario asked, "I don't know your name and I feel awkward to ask you for a date."

The young lady pressed his temple with soft fingers and drew them back towards her, "My name is Linda and you will not have time for dating, so don't ask."

"I'll be on a war bond drive. Perhaps our trails will cross at some point."

"Perhaps, but as soon as this plane lands, you will forget me. I won't blame you, we've just met and you will be soon surrounded by many admirers." She pressed his hand and gave him a telephone number. "Call me if you need anything."

As she walked up the aisle with a slight wiggle, the General observed, "Very nice young lady." Mario shrugged with approval and began reading the magazine.

The plane circled once over the airfield and made a smooth landing. The aircraft came to a halt and Mario dreaded the unfolding circus. The portable stairwell rolled up to the airline's door. Gen. Biggs led the way as Mario followed behind him and a military band began playing. The crowd in attendance screamed in approval. Mario walked slowly down the stairs and saw his mother and father, Hogan, Patrick, Benny, Emilio along with the rest of Mario's relatives. The Attorney General was present with a contingient of politicians.

The band played on as Mario embraced his parents. The crowds were held back by military police who formed a corridor leading to the motorcade. Mario hugged his Uncle Hogan, too. The old

Stateside Welcomes The Hero

man was crying. Patrick Morgan shook his hand and embraced him. The commander also had tears as he said, "You did it Mario, beyond everyone's dreams. You're the toast of America."

The Attorney General shook his hand and said, "We are all so proud of you." People in the crowd began throwing flowers at the hero as he returned the compliments waving his hand. Mario turned to both Patrick and the Attorney General and spoke over the din of the crowd, "I thank both of you people for having faith in me. I've thought of you constantly and believe it or not, it helped me during critical times." He told the Attorney General, "If there is anything I can do for you now or in the future, I'm at your call."

They reached the limousines as flowers continued to rain on Mario and the entourage. Mario, his parents, Hogan, Patrick and his wife and the general rode in the same limousine. The other members of the group crowded into other cars to complete the motorcade. A police motorcade escorted them to the White House.

Mario sat back and tried to relax. He mockingly wiped his brow and said, "Boy, I wouldn't want to go through that again."

After introducing the General to everyone in the limousine, his mother said, "This is nothing. Wait until you get home."

"There's going to be a parade like they have on the Fourth of July. Just like the one in Bristol," Rocco added.

"I think it will be easier for Mario to face the President than what awaits him in his hometown," the general said.

"I'm really not nervous in meeting him and receiving the award. It's the state dinner that makes me uneasy. I hope they don't ask me to give a speech," Mario replied.

Gen. Biggs said, "Don't worry about it, son, I'll do the talking."

Mario laughed, "That's what I'm worried about. If you do the talking, I'm going to be embarrassed. I don't want any praise."

"Did you get a chance to go to Pico?" Rocco asked.

"Yes, Pop, and everything is fine. Your brothers are settling down and Mama was constantly asking about you."

"How is she?"

"Don't forget, Pop, she is old and always called me Rocco, but on the day that I left, she called me Mario. I'll tell you all about my visit when we get home." Gen. Biggs raised his eyebrows and hummed a tune.

The motorcade entered the White House grounds and was directed by military police to a designated area. They all got out

Pico

of the limousine and other cars and were escorted to the Rose Garden section. The President came out to greet them. He shook Mario's hand, embraced him and declared, "Meeting you is a great pleasure. If there is anything I can do for you, name it."

"Sir, meeting the President of the United States is the greatest honor that I'll ever have and I have one request, sir."

He placed an ear close to Mario and cupped it, "What is it, Sergeant?"

"Authorize another star for my good friend General Biggs here. He is worthy of having a combat division."

The General's face reddened. Mario embarrassed him for a change. The president shook the General's hand and replied, "I'll see what I can do." He turned and gave a sly wink to Mario. They all had the opportunity to shake the hand of the President. He gave Mario's mother a kiss on the cheek and said, "America thanks you for having such a courageous boy in this world."

A presidential aide interrupted the gathering and announced, "We are ready for the ceremony." The President excused himself and left. Mario and Gen. Biggs were led to another area.

The rest of the group ambled to the front rows of the presentation area. The band was ready to play on cue. The elite honor guard of the military stood at attention.

The military band began playing "Hail to the Chief" as the introduction was about to start.

The President walked out to the podium to a round of applause. He raised his right hand in Mario's direction and waved him forward. The Sergeant moved with military precision to the delighted roar of the friendly crowd. The President happily spoke into a microphone as newsmen began taking notes and photographers beamed their cameras unto Mario. "This is your day, Mario, bathe in its glory."

The crowd replied with applause. Mario saluted the Commander-in-Chief who gracefully returned the gesture.

The crowd mellowed down as the President began the ceremony.

He pulled a folded paper from his jacket pocket and began to read, "The American that we are honoring today risked his life, not once, but on two different occasions. Each feat of bravery, above and beyond the call of duty, is an exemplary military accomplishment by Sergeant Mario Calcagni and certainly merits the highest military honor.

Stateside Welcomes The Hero

He saved thousands of lives while risking his life by directing artillery fire to within twenty yards of his position. Prior to the artillery barrage, Sergeant Calcagni was in hand-to-hand combat. When the smoke cleared and the battle later analyzed, this great American soldier stopped a major counterattack that, had it been successful, was calculated to doom an entire Division. The parents and loved ones of those men are eternally grateful."

The President took the first blue-ribboned Medal of Honor and placed it around the sergeant's neck. The crowd was awed in silence. The President then took the second Medal and smiled as he placed it around his neck. The President then stepped back and saluted the American hero. Mario returned the salute as the President said, "Thank you."

Mario responded briefly, "Thank you for the opportunity."

The ceremonies were over. The crowd gave Mario a standing ovation. Photographs were taken profusely. Movie cameras were rolling as the President shook hands with Mario and gracefully departed with a victory wave. The crowd milled about as newsmen began interviewing Mario and the General. A reporter asked Mario, "How does it feel to be a hero?"

He thought for a moment. "I feel fine now but at the time, I was probably temporarily out of my mind."

Another asked, "Were you scared?"

"Yes, always."

The General smiled as a magazine writer asked Mario, "What are you going to do now that you're home?"

"The first thing that I'm going to do is go to church and thank the Almighty for bringing me here safe and sound. Then, a little vacation and later go on tour to sell war bonds." Mario stepped back and introduced the general with a hand gesture and graceful bow. The reporters responded with laughter.

Gen. Biggs was asked, "What do you think about that man standing beside you?"

"What he did was beyond words. I am here today because of him."

"Did he realize then that he was saving the life of a colonel?"

"No, it was impossible for him to know. We were not even in the same division and it was pitch-dark when he found me."

One reporter asked, "General, I guess you had a ringside position."

Pico

"It couldn't be any closer. We were in a ring and Mario was between the rest of us and the enemy." Gen. Biggs excused himself and Mario. The reporters gave them a courteous round of applause.

The Calcagni entourage was escorted to their hotel. They had a few hours to relax before attending the presidential dinner. Mario exchanged stories with family and friends but didn't go into any details. He enjoyed the pieces of information about people in his hometown.

The dinner that night was a formal affair and the Calcagni group was visibly uncomfortable. The event was a new experience for them. Afterwards, they were happy to get back into the confines of the hotel. The next morning, they would all board a train for home. The general was to accompany them. It appeared that his designation, for the time being, could best be described as Public Relations Officer.

Hogan tried to get the General to shoot a friendly game of pool. Mario gave his uncle a dirty look and he dropped the proposal. It was two in the morning when they all retired to their assigned rooms. Mario was in a suite with his parents along with Uncle Hogan. He laid back in his bed and realized that the difficult part was over.

He had met the President along with the dignitaries. The awards ceremony was completed. The rest of the program that laid ahead should be a piece of cake. Nevertheless, he had difficulty sleeping.

The next morning, everyone arose to an early breakfast and were later escorted to the railroad station. The trip back home would take five hours and the entourage was accompanied by numerous reporters and cameramen. The entire trip was a festival. Passengers danced and sang in the aisles.

Benny thought that the young ladies on the train were all after Mario, who evaded most of them. The General said, "Mario is going to be on the cover of some national magazines and is already on the front pages of most daily newspapers."

"I heard this train is normally at half-capacity. The seats were sold out last night," Patrick said.

Mario thought the worst was over. It was most difficult to negotiate a trip to the restroom. Benny and Emilio laughed as Rocco expressed some concern for his son's welfare. He said, "Mario survived war with the Germans but these screaming girls are going are going to drive him crazy." His son returned from the restroom, his tie and hat were missing and his hair was messy.

"What the hell happened to you?" Hogan asked.

"Somebody better do something to control these goddam broads before they kill me."

Benny laughed hysterically, "I never thought I'd see the day that you would be running from girls."

Mario smiled as a beautiful brunette broke through the crowded aisles and tried to kiss the hero. Mario gently lifted her in the air and directed her to Benny. He told her, "Kiss him, he won a Silver Star."

It was a rowdy train ride until the general enlisted the aid of some military passengers to maintain some semblance of order. They would soon participate in the upcoming parade festivities that was to begin at the railroad station.

The train pulled into the station as the crowd waved flags and screaming for their hero. A military escort was provided to protect Mario.

The crowd was tumultuous as the hero stepped into a path cleared by the military guard. Girls pelted him with roses and others tried to touch him.

He smiled, waved and at times, tried to shake hands but the crowd was riotous. There appeared to be no respite in the frenzied celebration. Mario and his party were rushed out to a secured area and jumped into open convertible limousines. His family, Patrick and the general rode in the same car. One of the several bands played up as the lead car joined the procession and the parade started.

Mario sat alone on the rear seat. The public wanted to see him. As the joyous procession slowly moved along, Mario beamed to the happy cries of the crowd. He waved and blew kisses. Within a half-hour, the car was littered with roses. His parents were smiling from ear to ear. Mario was all they had and he was home. They were proud of him.

Mario saw his old nemesis, the parish priest, in the crowd who blessed the car as Mario ordered the driver to stop. He waved the priest over and helped him into the open limo. The crowd applauded in approval. The General told Mario, "You sure have a lot of friends."

Pico

"Yes, I really do have a lot of friends, but not this many."

The parade entered the high school stadium where the politicians were waiting. The governor was in attendance, senators, along with congressmen and the mayor. They all made speeches and bathe in the reflected glory of their homebred hero. Mario was called upon to speak as the mass of admirers began applauding.

He waved in consent and came to the microphone. When the crowd finally quieted down, he started out:

"I want to thank all of you for the warm welcome. I appreciate it but it wasn't necessary. I was fortunate to be in the right place at the right time and I acted in the best way I knew how. I feel awkward with all this adulation and ask that I be treated as one would treat a good neighbor. I would now like to call on the good Father here to lead us in prayer for the other brave men, who, unfortunatly will not be coming back to their home. Thank you and God bless you all. You're the greatest people in the world." He waved to an ovation. The crowd was silent as the priest began the prayer.

Mario and the family finally made it home. The General soon departed for his hometown. He left a military escort to assure Mario's family some privacy. When he left, there were still over fifty observers keeping company. Some were reporters and others were admirers. Mario laid back on the couch and let out a long sigh, "It's all over, I dreaded these few days. From the time I was informed of the awards, I knew the consequences. Boy, what a load off my back."

His mother embraced him and said, "My poor Mario."

"Really, I mean it, this is the very first time that I'm happy." Rocco sat proudly next to his famous son. He had a glass of wine and asked Mario if he wanted some. His son yawned and stretched and replied, "No thanks, Pop, I'm going to be like my paisanos in Pico and have wine only when I eat."

"Help yourself whenever you want, I made two barrels of zinfandel this year."

"Pop, do you know anything about the Micheletti family who lives somewhere here in New England?"

"Of course. They live in northern Massachusetts. You may remember them. They visited here years ago when most of his family came from Pico."

"Angelo's wife refused to come to America. She and her daughter remained in Pico," Carmela said.

Stateside Welcomes The Hero

"There are two daughters back there. Anna raised her niece, Gabriella from birth," Mario said.

"Nevertheless, her place should have been with her husband and other children. They are a family divided," Carmela said.

Rocco stared at Carmela and said, "Mind your own business."

"Her parents were old and Anna didn't want to leave them," Mario explained.

"Her place is still with her husband and she should be prepared to come to America."

Rocco took a sip of wine, "Women can't mind their own business. Why are you interested, son?"

"They are great people and are very close to our family in Pico. During the most critical days of the war, they took care of your mother."

"Where was Giovanni and Luigi?" Rocco asked.

"Its a long story, they took their families into the mountains to hide and Mama was too old to move around the rough terrain."

"Do you feel like talking about your visit to Pico?" Rocco asked.

"Sure, why not?"

Carmela sat with them and Mario related his experiences but excluded most of the atrocities and killings. He spoke only of the good times with great embellishment.

Carmela said with concern, "Your general indicated that you were awarded three Purple Hearts. Where were you wounded and are you all right?"

"I'm fine. The first two weren't bad at all. They were only shrapnel wounds. The last one occurred less than a month ago. It happened in Peter Colombo's house."

"What happened?" Rocco asked with concern.

"I was shot in the leg, but again, I was lucky. The bullet went right through. No problem, just a memorable scar."

"No more war for you!" Carmela declared.

"Don't worry about a thing. Even if I wanted to go back, they won't allow it and I'm in perfect health now."

"How is Peter?"

"Mr. Colombo was dead before I had a chance to meet him. The Goumiers killed him and his entire family."

Carmela raised a towel to her face and cried, "That poor man, he worked so hard. He should never have gone back to Italy."

Pico

Rocco shook his head. "Your mother is right, he should have stayed here."

"Some renegade Goumiers were using his farmhouse as a hangout. With the help of Giovanni, I cleared the property."

Carmela said, "You should have been more careful. You mean to tell me that they listened to you when you told them to leave?"

"Mom, it didn't happen that way. Someday, when you and Pop visit Italy, Giovanni will tell you all about it."

"I'll never go back. If your father wants to go, he can go, alone."

Mario replied with firmness, "Someday in the future, you are both going back together for a visit. Pop, when things cool down, I would like to visit the American side of the Micheletti family. I have much information to share with them."

"Whenever you say, they live about ninety miles from here."

Carmela said, "Of course your father will be happy to visit them. Angelo makes wine every year." Mario laughed while patting his father on the shoulder.

Carmela continued, "Perhaps Anna and her two daughters will move to America when the conditions allow."

"I don't think so, she has a beautiful piece of property. If Angelo wants his wife, he better think of relocating back there," Mario replied.

"Are you going to suggest that to him?" Rocco asked.

"Only if he broaches the subject. Gabriella is twenty-one years old and she will never come here. Besides, she is going to marry a German who is going to live there."

Carmela looked up in amazement, "What do you mean, a German?"

"I met him, he is a soldier who wanted to quit the Army and run off into the hills with Gabriella. They are in love with each other."

Carmela laughed, "He will be back in Germany after the war and she will never see him again."

"Not true, Mom, he is in Pico right now and my guess is that they are probably already married."

"How do you know so much about this German?" she asked.

"He is my friend and I have a lot of respect for him." Carmela was evaluating over the possibility and suggested for the sake of conversation, "Anna should give the property to Gabriella and the German. She and her younger daughter can join their family here."

Stateside Welcomes The Hero

"You have a good point. Anna's mother and father were both victims of the war but there is one very big problem."

"What can be so important?"

"Little Angelina, she will never depart from her beloved mountains."

"It is their problem to unravel," Rocco said.

Carmela replaced the telephone back on the receiver and it rang immediately. She answered and Benny asked to speak to Mario.

"What can I do for you?" Mario asked.

"Can we meet sometime tomorrow? I need to talk to you!"

"Not tomorrow, but the day after will be fine. My house here at about nine in the morning."

Benny replied happily, "You got a deal, see you then."

Mario stretched and announced, "I think I better hit the sack. Tomorrow morning, Patrick and I are going to pay a visit to the Attorney General. I never had a chance to thank him because of all the commotion."

Everything was as he had left it as he fell soundly asleep. He was home safely with his family. The next morning was bright as birds sang in the nearby trees. He showered as his mother prepared breakfast.

Mario sat with them and said, "It is the best night's sleep I've had since the night before I left Pico."

"Why's that?," inquired Rocco

"My two uncles were there and I felt protected."

Carmela left the telephone off the hook for the entire night. In fact, it was still inoperative. There was a knock on the door and Patrick Morgan entered. Rocco had him sit down to a cup of coffee as Mario went to his room to don his military uniform. He called out to the commander, "How are we getting there?"

"I have my car out front. We will get one of the military guys to drive us. That's what they are getting paid for."

"Let's get going."

"There are still some reporters outside. They will probably follow us around," Patrick said.

"No problem at all, they will get a tour of the State House."

There were three Army security guards present as Mario asked, "Anyone here drive a car?" They could all drive and Mario selected one, "Okay, what's your name?"

Pico

"Jeff and I are from California."

Mario pointed to the car as Patrick handed him the keys. He told the other two to continue whatever it was that they were supposed to do.

As they approached the car, a few reporters and newsmen asked some questions. Mario answered courteously and excused himself. One inquired, "Where are you going?"

"We are going to the State House to see the Attorney General." They sped away as Mario gave directions. The reporters followed. Jeff was driving fast as Mario thought: I assume they all drive fast in California.

Shortly, they heard the sound of a siren and were directed to stop. The patrolman approached the car with his pad open in hand. He began reading a citation to the military driver when he looked in the back seat and recognized Mario. He stopped in mid-sentence with his mouth half-open and asked, "Aren't you Mario Calcagni?"

"Yes, I am."

The policeman handed him his ticket pad and asked for his autograph. Mario signed it with a smile and handed it back to him. The officer offered to escort them to their destination. Mario thanked him and explained that they were not in a hurry. The policeman waved them ahead along with the reporters.

"Is your brother-in-law expecting us?" Mario asked.

"No, but it doesn't matter. You don't need an appointment."

Mario patted the Irishman on the knee and asked with concern, "How are you feeling, Patrick?"

"I quit smoking and have become a vegetarian. Seriously, I do feel good. Sure, I have a heart problem but I can't complain."

"Good, my friend, I am happy for you. Are you still eating that soft-boiled egg every morning?"

"No, believe it or not, the doctor told me that eggs are bad for people with heart conditions."

Mario laughed, "That clown must be nuts, I always heard that eggs were nutritional."

"Regardless, I eat rye toast with a cup of coffee."

"You're telling me that you only drink one cup of coffee a day?"

Patrick laughed slyly, "I'm only supposed to have one cup a day."

"How many do you have?"

Stateside Welcomes The Hero

"None of your business." He changed the subject, "Did you get a chance to try out some of those Italian girls around Naples?"

"Pat, don't try to change the subject, I'm going to tell your wife about your interest in other women."

"Go right ahead, she won't believe you."

They pulled up to a State House parking space. The reporters did the same and followed them into the capitol building. They asked Mario as they walked in the echoing hallway, "Why are you seeing the Attorney General?"

"He is my friend and I want to visit with him." They entered a large outer office and the staff of girls were screaming with delight. There was a time, not too long ago, when he would become embarrassed but he was getting accustomed to the display of recognition and emotion. He signed some autographs as the receptionist ushered them into the Attorney General's office. The Irishman came around his desk and knew from the disturbance that Mario had arrived.

The hero shook his hand and explained, "I didn't have much of an opportunity to properly thank you. For the past few days, I've been herded from one location to another, but I'm here now."

The Attorney General replied, "My friend, there is no need to thank me. You have performed in a manner that is worthy of you. We are all indebted."

"No matter how you cut the cake, Patrick got the ball rolling and you gave me the opportunity." Mario took his Purple Heart ribbon and pinned it on the blind man's jacket. He shook his hand and said, "Thanks for giving me the opportunity."

Patrick explained to the sightless official, "He gave you his Purple Heart."

The Attorney General wiped away his tears and regained his composure. He asked Mario, "Have you considered running for political office?"

"No, I don't intend to but if you ever decide to run for Governor or the United States Senate, I will campaign for you."

"Thanks, son, but I believe this is as far as I can go."

"Hogwash. I'll let you in on a little secret. Whenever I was faced with an impossible situation I thought of the Attorney General of my home state who is blind. I said to myself: That takes real courage. To be a high law authority in the state without being able

Pico

to see an adversary. My situation took less courage because I had all the tools at hand, including being able to see the enemy."

"Thank you, Mario, with you behind me I fear nothing. When the time is right, perhaps I will run for governor."

"Sir, you've never had any fear of anything." They embraced. Mario and Patrick left the office soon to be confronted by a throng of reporters. They asked, "What's going on Mario?"

"Nothing much except that the Attorney General may someday run for governor." They drove back to Mario's house. The ex-navy commander playfully saluted and said, "I'll see you later."

Mario found the house full of friends. They were all there, Hogan, Emilio and all the relatives. Everyone was drinking wine along with a feast of spaghettini con olio e aglio. Rocco sat at the middle of the table and directed matters. He commanded Carmela, "Get some more bread. We need more wine and glasses. Are the wild mushrooms cooked yet?" He was in his element. The military escorts were in the house and he was getting them drunk. Mario shook his head and thought to himself: Just like old times.

It was nearly eight in the evening when they finished drinking coffee royal. Everyone left except for Hogan. Carmela washed the dishes as Rocco, Hogan and Mario were eating mixed nuts. The telephone was still off the hook.

There was a knock on the door. A telegram addressed to Mario. It was from General Biggs explaining that he tried to call but the line was always busy. He instructed Mario to call him. It was a matter of importance.

Mario said, "What now?"

He went to the telephone and dialed a long distance number somewhere in Michigan.

Gen. Biggs answered, "What's going on out there, I can't get through?"

"Too many calls, we had to leave the phone off the hook."

Biggs explained, "I've got one hell of a deal for you. The breakfast cereal people out here want to use your photograph on their cereal boxes and they will pay you a royalty for your endorsement of their product."

"No kidding, is it a good deal?"

"Of course. Good cereal and they need heroes for the kids."

"If it meets your approval, it's fine with me."

Stateside Welcomes The Hero

"Fine, I'll give them the go-ahead and they will be contacting you soon. Give my best to your family and I will see you in a few weeks."

Mario hung up and explained his good fortune. "That is a clean operation that I'm happy to be involved with."

"How much money are you going to make?" Hogan asked.

"I don't know, it depends on how many boxes of cereal they sell. By the way Hogan, I visited your home village in Italy. They didn't even have a pool room there. In fact, there were hardly any people around."

Hogan replied, "There weren't many to start out with. After I left, the place went to hell." Rocco shook his head in disbelief and thought Hogan was crazy. Mario went to bed as Rocco and Hogan continued on for another hour.

25

THE GREAT PIZZA AND PASTA VENTURE

The next morning, Benny arrived on time. They had coffee and went into the living room. Benny said, "We need to find time to get you laid."

Mario was receptive, "Yes, that's a good idea, but, I know you didn't come here to tell me that. What's on your mind?"

"Let's go for a drive. I want to show you something."

"Sure, why not?"

Benny injected casually, "Some reporters are still outside and they might follow us."

"Fuck the reporters, we'll lose them and have a little fun." Mario put on civilian clothes and said, "Let's go."

Benny's white Cadillac appeared to be in immaculate condition. Mario looked the car over and said, "It looks like you are doing great, congratulations."

"Anyone that hasn't gone into the military is doing well. All the overtime one can handle. I have a connection for getting gas rationing stamps and I'm making money hand over fist."

"No kidding? How much money are you talking about?"

Benny beamed, "Would you believe over one hundred grand?"

"Wow! While I am getting my ass shot at you're raking in the fucking money."

"Mario, you're only partly right."

"What do you mean by that bullshit?"

Benny had lost the few reporters that were following them and replied, "With your name and my money, we are going to hit it big time."

The Great Pizza and Pasta Venture

"Yeah, how do you account for all the money?"

"My wife has been working and my trucks are kept busy all the time and besides, who gives a fuck about where money is coming from? Everybody is doing great."

"Benny, I went through hell to clear my name, do you think I want to throw it all away?"

"You would not be taking any risks whatsoever. As soon as you approve of what I'm proposing, I stop the stamp business. I'll be as clean as a whistle." Benny drove to the Italian sector located on one of the hills in the area.

It began raining as they stopped on a deserted lot off a main shopping area. The side streets were already swarming with abandoned shopping carts. The rain intensified as they sat in the car. Benny pointed in the direction of an old one-story brick building. He explained enthusiastically, "I bought this piece of property for less than five thousand dollars."

Mario laughed, "You sure screwed yourself, this whole area is nothing but trouble."

"No I didn't. Things are changing fast here and all for the good. Come on, I'll show you the inside of the building."

"What in the hell do you have in mind for this rat's nest?"

"The best Italian restaurant in all of New England." Mario laughed, "I knew you were always crazy but this takes the cake." Benny didn't pay any attention, "Come on, let's go."

The door, which resembled a garage entrance had a big padlock on it. Benny fumbled with a bunch of keys. The rain was coming down harder now, "How are the broads in Naples?" he asked.

"Is that all you have on your mind? Get the damn key in the lock."

"I think I have the right key, but it won't go into the lock."

Mario gave some key advice, "Put some hair around it."

Benny finally got the lock open and the old door creaked as they entered. The building reeked with a musty odor and there wasn't any lighting. Benny lit an old kerosene lamp and Mario saw the interior was littered with junk. He weaved his way through until he reached the far end of the large structure.

Mario thought to himself: What a waste of money.

His friend held the lantern high and said, "Look at that." There were six large brick ovens that extended twenty feet in length.

Pico

"Benny, how much Italian food are you going to bake in these enormous ovens?"

"A lot. We can also utilize them to bake bread."

"Don't be so pessimistic."

"Me pessimistic? Italian food is probably seventy-percent pasta and most pasta is boiled. What the fuck do you intend on doing, boil the pasta in wood or coal-fired ovens?" Benny lowered the lantern. "Let me show you the cellar." They went down a set of creaky wooden stairs and stepped off onto a dirt floor.

"Who sold you this white elephant?" Mario asked.

"The Don and he said because the building is in such bad shape, I can do whatever I want to do with it."

"Why do you need permission from him?"

"He owns this territory. Everything here goes through him. He doesn't want a restaurant on every corner and operates the area the same way as the state controls the number of liquor stores. The man is fair."

Mario smiled, "Why did he reward you with such a sweetheart deal?"

"Because I took this broken down building off his hands."

Mario laughed, "The Don doesn't need anyone to take anything off his hands. There is more to this than meets the eye." Mario pounded the dirt floor with his foot and joked, "I wonder how many people are buried here. This place is inhabited by ghosts."

Benny raised the lantern for more light, "Look at the size of this cellar. Some of the space can be used to store wine. I tell you, this place has tremendous possibilities."

Mario followed Benny up the stairs and asked pointedly, "Did you ever indicate to the Don that I would be involved with you on this venture?"

Benny hesitated and stuttered, "I told him that we are good friends."

"That's great. I told you that I don't want to get involved with anything crooked. You really surprise me, pulling this shit on me."

"What the hell are you upset about? I only bought a broken down piece of property. Besides, he respects you," Bennie responded.

"That is not the issue, I like the man. The problem is guilt by association."

The Great Pizza and Pasta Venture

Benny retorted, "You're are creating obstacles. There is nothing wrong with buying this property and I paid the fair market price."

"Benny, use your head. All these reporters are following me around. Can't you just see the headlines?"

The little entrepreneur replied, "It is in my name and we can leave it that way. There is no connection. Isn't it possible for Italians to open an Italian restaurant in little Italy?"

Mario thought for awhile as Benny kept his fingers crossed. "Presented that way, it could fly but only if the establishment is operated above and beyond reproach."

"Of course it will but I don't know what the word reproach means," replied Benny.

"It means clean and I emphasize, squeaky clean. I'm not giving my consent yet but I mean it, no running against the law."

It was still raining when the door opened and three men entered. One of them called out, "Who the fuck is in here?"

"It's me, Benny."

The Don knew beforehand, the Cadillac outside gave him away. He was only pulling Benny's leg. Johnny Catalano and a huge man accompanied him. The Don saw Mario with Benny. He came forward and embraced Mario, "Thank you for everything you've done in Italy. You know, they are going to name the main road outside after you." Mario said, "I haven't heard anything about it."

The Don placed an arm around him and replied, "I'm privileged to inside information."

"You eliminated over forty Moroccan Goumiers in the Pico area and some of the people you helped were my relatives. The bastards you killed in Colombo's farmhouse were the same ones that committed atrocities in my old village. I'm indebted to you."

"How did you get this information, I haven't told anyone?"

"It's irrelevent but like I said, I am privileged to all sorts of information." Johnny Catalano stuck out his hand to shake Mario's who didn't know what to expect.

"Now, I don't feel so bad about the incident on the bridge. You enhanced my reputation," Catalano said.

The Padrino looked at Cats with surprise, "Enhanced? What happened, Johnny, did you swallow a dictionary?" Everybody laughed, but Johnny was embarassed and forced a grin.

"Don't worry about association. I won't ever come into your restaurant unless I'm invited," the Don said.

Pico

"If I decide to join up with Benny here, you have a standing invitation, Mario assured."

"Thank you, I appreciate that. If you ever need anything, feel free to call on me."

As the three men left Benny said, "I told you, there are no problems whatsoever."

Meanwhile, outside Johnny told the Padrino, "That fucking Mario owes me and his day is coming!"

The Padrino looked at him with piercing eyes, "Mario is my friend and I like him. Forget the past. He's untouchable and doesn't concern you."

"I have a score to settle with him," Johnny replied.

The Padrino was patient but spoke sternly with finality, "Perhaps I didn't make myself clear, there will be no confrontation between you and Mario. Do I make myself understood?"

Johnny cleared his throat, "Yes, I understand."

"Good, Johnny, in spite of your beastly instincts, you listen to my good advice."

The Padrino was excellent in judging character. It was part of his business. However, he knew that Johnny Catalano would ultimately act as he so desired.

Meanwhile, Mario and Benny were contemplating the last words from the Padrino. Mario said, "I guess you're right. I never heard of the Don going back on his word."

"He has the greatest respect for you and you know the meaning of that word." Benny responded.

"Benny, sit down. About this operation I think I have an idea."

"I'm listening with both ears." Benny was happy that his friend was showing some interest.

"Naples is more popular for its excellent food than for its beautiful and enticing women. It is my opinion the future trend in Italian restaurants in this country will be inspired by Naples' cuisine."

Benny listened as if Mario was an expert. "Many of the restaurants there are pizzerias. They eat pizza there like we eat hot dogs."

Benny said, "I believe it. Some of the local Italian bakeries sell pizza whenever they have extra time after baking bread. Otherwise, the only way you can get a pizza is if your mother or grandmother bakes it."

The Great Pizza and Pasta Venture

"That's right Benny. But in Italy, specifically in the southern cities, it's available on every corner and you know what?"

"What?" Benny asked.

"All these soldiers who will be returning from Italy after the war will be craving for pizza. I've seen them when they were on leave. All they ate was pizza. I'll let you in on some other information. The Neapolitans bake pizza in ovens that are exactly like these!"

Benny replied with enthusiasm, "Wow, we're sitting on an idea that could be a gold mine. I can see it now. He waved his hand and said, "Calcagnis' Pizza and Pasta Grotto."

Mario smiled with guarded approval. "In the event that you want to carry pasta I recommend it. In Italy most restaurants make their own pasta. We must do the same. The objective, at all times and with every product, must be to offer nutrition without sacrificing taste."

Benny scratched his head, "How did you learn so much in such a short time?"

"It's easy, when one has an interest in something, the absorption factor is phenomenal. Before I got into the war zone, I thought about what I would do when I returned home. Most of the people around me talked about buying a gas station. Me, I visualized an Italian restaurant. Naples was on hand and I understood and could interpret the Italian cuisine. They say that Italy is the art museum of western civilization. I agree and will go further and say Italy is one of the best culinary capitals of the world."

Benny shifted in a old beat-up chair, "I am impressed. Your name alone will make us wealthy, but your ideas will assure success beyond what I ever could imagine. I put up all the money and we will be fifty-fifty partners."

Mario shook his hand. "You've got a deal. But, I don't want you to borrow money from the Don or his associates. Time is of the essence, the building must be renovated and opened within six months. We must capitalize on this hero aspect while the fervor is still on."

"Mario, we will be famous," Benny added.

"That's what I'm worried about. By the way, plan on someone operating a pasta machine for public view while customers wait to be seated."

Pico

"Great, great idea."

Mario was on a roll, "Be sure to employ an Italian architect who can redesign the building to adopt the old-world style. That is very important. Veterans will tell family and friends this is exactly as it looked in Italy."

It stopped raining and the sun broke through the clouds. They stepped out of the building and went to the car. Benny made a right turn and sped along the street that was soon to be designated as Calcagni Avenue. Mario asked, "Where the hell are we now headed?"

"I told you I was going to get you a great piece of ass."

"Don't be ridiculous. If the word ever got out, it would hit every newspaper around," Mario argued.

"Leave it to me. I've been planning this for some time. Nobody is going to see us and the girls know the score."

Benny drove into an underground garage of a secluded apartment complex. "How do you know that these girls will keep their mouths shut?" Mario asked.

"I know for a fact because the Don uses them and for friends."

"Who is paying for the pleasure?"

"Me. It's a homecoming present and these girls are yours anytime you desire."

The hookers were exceptional and experienced. Mario had a good time and when he exited the bedroom said, "Not bad, not bad at all."

"I knew you would enjoy yourself. Where do you want to go next?"

"Maybe a little lunch before we head back home."

"I know a place near the racetrack that recently opened. They serve prime steaks that are over an inch thick."

"How do you know that the meat is prime?"

"Well, it's not all prime but special people get it. The rest of the customers pay for prime, but get choice."

"Don't tell me, you qualify for the prime category," Mario asked.

A driver cut in front of Benny who swerved and yelled out, "Va fongoo." He told Mario, "Yeah, I give the owner a few gas stamps and I get the prime cuts."

"Benny, the days of stamps are history."

The Great Pizza and Pasta Venture

"Yes, today is the last day. I'm going to turn the operation over to Emilio."

The owner was working in the kitchen as Benny tapped him on the shoulder then tried to hide himself. The burly chef, Nick, turned around. He playfully flashed a butchers knife and said, "Stop breaking balls."

"Broil two of your best steaks for Mario and me and a bottle of your private wine."

Nick looked at Mario, "You're the local hero, aren't you?"

"I'm Mario Calcagni."

Nick placed the knife on a nearby table and wiped his hand on his apron. He shook Mario's hand and said, "This is a pleasure, you are my guest." He called out to his wife and daughter.

His wife complained in Italian, "What do you want? Do I have to hold your hand every minute of the day?" She entered the kitchen and immediately recognized Mario. Her daughter closely behind let out a scream of joy.

Nick introduced them, "This is my wife Olga and my daughter Santina and I guess I don't have to tell you who Mario is."

They took turns in embracing an embarassed Mario. The women knew Mario's discomfort and said, "Not here. We will arrange a quiet secluded booth that will give you privacy." Santina was beautiful and Mario asked Benny if she was married.

"Why are you asking me? Ask her yourself or do you want me to beat your time."

Mario moved into the booth and laughed, "You could never beat my time, not even with your sister-in-law. Have you fucked her yet?"

"No comment."

"That means, either you haven't or you got caught."

Santina presented the homemade wine and explained, "We don't sell this, it's too precious."

Mario said, "Just like you. Are you married?"

Her face reddened and smiled, "No, I'm not."

She bent over them and poured the wine displaying a little cleavage.

"Mama Mia," Mario uttered.

Santina quickly stood upright. "I'll be back with the salads."

"Aren't the salads served last?" Mario asked as Santina gave a perplexed smile.

Pico

Benny said, "Not a bad place. It holds about fifty customers. Our place will serve about two hundred."

The girl brought the salads with plenty of fresh Italian bread. She poured more wine. "Your steaks will be ready in a few minutes."

The restaurant served basic Italian food and Mario noticed that nobody was eating steak. People occasionally stared in his direction but those who recognized him respected his privacy. He wasn't wearing his uniform so most of the customers weren't sure.

Santina brought the steaks and Mario said, "I hope the steaks are good because we are the only ones, that I notice who ordered them."

She replied playfully, "That's because steak isn't on the menu."

She walked away and Mario queried, "Why do we get steaks in a restaurant that doesn't list it on the menu?"

"Don't worry about it. The meat is stolen. A few stamps here and a few stamps there and anything is available," Benny replied.

Mario responded as he placed a morsel of meat into his mouth, "You've come a long way, where did you go wrong?"

"I didn't go wrong, I only took an opportunity that was presented to me."

"This steak is fantastic. Where in the hell did they steal it?"

"I don't know the specifics, only that it was supposed to go to a quality hotel for a big political affair. Those thieving bastards don't even need stamps for meat, let alone this quality. Somebody up the hill heard about it and intercepted the shipment. The rest is history."

"Did you know anything about it?"

"I heard about it and that is why we are here enjoying the very best."

"What did the guests eat at their gala event?"

"Who gives a fuck? But I heard they dined on hamburger," Benny replied.

"This place lacks music, make a mental note. We will offer piped-in continuous music in our place," Mario observed.

Benny was happy he had Mario as a partner.

Mario enjoyed the meal and reminded Benny, "Today is your last day with the stamp business and anything else you have going that is illegal. I hope you understand because if I hear otherwise, the deal is off and I will break your fucking neck."

"Hey, don't worry, kid, I give you my word. I'm going straight."

Mario gave Santina a kiss on the cheek and promised, "I'll call you."

As they walked through the kitchen. Mario thanked Nick and his wife and they walked out the back door. Nick turned to his wife and advised, "We should get a large autographed photograph of him and display it in the dining room."

Benny drove the car onto the highway without looking for oncoming traffic. Mario shook his head, "You're going to get us killed."

"No way, there are very few drivers out here."

Mario was playing with him, "Why?"

"They don't have any stamps and what gas they have they save for Sunday."

Mario asked, "How is my horse Sparky? I haven't thought about him."

"The fucker died. It must have been out of loneliness. You left and the animal refused to eat. The old man didn't want to tell you. and my hedges are beautiful now."

Mario tapped him on the side of the head and said, "Fuck your hedges."

Benny drove up to the house. The military guys were still in there. Mario got out the car as Benny revved the engine. He reminded him, "Don't forget, I'm serious. No more screwing around with illegal business."

Benny saluted in approval and said, "I'll have some plans ready for your review in a couple of weeks." He raced away.

Mario told the soldiers, "Don't worry, you'll be out of here in a few days."

"Don't rush it, we love Mama's food," they replied.

26

SANTINA'S LOVE FOR A LEGEND

"Hello, is this the lovely Santina?"
"Yes, Mario, this is Santina"
"Then I guess you remember me."
"How could I not remember our famous hero?"
"What are you doing tonight?"
"I just took a bath and I'm setting my hair."
"I guess you're going out. I'm sorry that I bothered you. Perhaps some other time?"
Santina sensed that he was ready to hang up and quickly said, "I always set my hair for Sunday Mass."
"That's tomorrow. How about dinner tonight? Of course, if you have previous plans, we can make it some other time."
"I had planned on going to confession."
"Go tomorrow morning before Mass, I have a special place in mind. Give me your address and I'll pick you up in an hour."
"Make it a little over an hour."
Mario hung up and called Benny. One of the children answered as Mario heard the other two fighting in the background. Mario asked, "I need the Caddie tonight, okay?"
"Why, kid, do you have a hot date?"
"Could be, the girl is going to confession in the morning instead of tonight. I could get lucky."
"Mario, they never confess what goes on during a date to a priest because the priest will ask them forever for the same accommodations."

"Stop breaking balls, you're out of line."

Benny laughed, "It's just like old times. Have fun with the broad and keep the caddie, I'm ready for a new one."

"Give me your asking price and I'll buy it."

"Yeah Mario, now, you're breaking balls. Have a good time. By the way, who is the girl?"

"The one we met at the restaurant."

"Madonna, she's built like a brick shithouse!" Benny replied.

Mario concurred, "That's an understatement."

It was a warm pleasant evening as the convertible stopped in front of the family cottage. Mario honked the horn and Santina promptly came out to the car. He opened the passenger door as she apologized for her unruly hair. Mario complimented, "The hair, like the rest of you, looks fabulous." He thought to himself; If this girl is a virgin, it won't be for long. She had smooth olive skin, wavy brown hair and flashing brown eyes. He looked her over and over and saw she was perfectly endowed.

Mario gunned the big car to the main highway as she asked, "Where are we going?"

"A friend told me about a beautiful new seafood restaurant on the Cape where you can order the lobster by the pound."

The breeze ruffled her hair as she faced the wind in her face, "I'm game, I love fresh lobster."

Mario was now driving at high speed as she exhilarated in the wind in her face and said, "This is like flying." Santina was on cloud nine. Everything was perfect. Above all, she was with Mario Calcagni. The feeling took hold the first time she read about him in the newspapers.

Mario explained, "The restaurant is called the Boat Landing. I hear that the building extends out over the bay. Lobsters are delivered by small fishing boats to a pier under the restaurant. They have a stairway from the kitchen to the pier."

"It sounds great. I'm looking forward to it."

"That's not all, they claim it takes only twenty minutes from the boats to the dinner table. I must admit, that's quite an accomplishment." Mario, caressed her knee, "Nice legs."

Santina smiled, "That is what I interpret as acting fresh."

The sky was clear and the stars sparkled as Santina asked, "Why isn't that star as bright as the others?"

Pico

Mario got a bit corny, "Because that is not a star. It is a planet. It's Venus and named after the goddess of beauty. I'll bet she looks like you."

Santina smiled at the compliment and enjoyed the view. She asked, "Tell me about some of your war experiences."

Mario hesitated, "I never like to talk about that. It's different and has no place in America as we know it. Let's talk about you."

She laughed, "I haven't gone anywhere or done anything special."

"Consider yourself fortunate. Many Russian women soldiers died fighting the Germans on the front lines. Some say they were as formidable as their male counterparts."

Her eyes were closed in imagination and awe, "I always felt that women could hold their own during battle. I wish I had the opportunity."

"Santina, most of them perished."

"Then, I take it all back, but life is so boring."

"Have you been out with many girls?" Satina asked.

"A few, why do you ask?"

"I hear through the grapevine that you have quite a reputation as a ladies' man."

Mario shrugged. "Grapevines are for grapes but tell me, what have you heard?"

She opened her eyes and continued to gaze, "I heard that if you don't score on your first date with a girl, you consider the date a failure."

Mario smiled and teased, "How do I stand with you, do I score or am I destined for failure?" He thought to himself: I screwed up. I'm moving too fast. She spoke softly, "Don't you have any respect for a good woman?"

"Of course I do. Girls are fun and I respect the opportunity for the good things in life."

"Mario, I'm only eighteen. The question you asked wasn't fair. What you call scoring is possible because you are who you are but I will do my utmost to avoid it."

"Are you saying that I have only a possible chance of scoring?"

"Why don't we change the subject?"

"You mentioned you go to confession Sunday mornings so what's the problem?"

Santina didn't reply. She hoped her silence was enough to end discussion on the matter.

Mario sensed that he was out of line but he decided to pursue the subject a bit further, "Okay, just one more question. If you don't reply, the answer is yes."

She looked at him curiously, "What kind of game are you playing?"

"No games. If the answer is yes then you keep quiet and I'll get off this sensitive subject."

She thought it over and replied, "Okay, ask your question."

"Are you a virgin?"

Santina laid her head proudly back on the seat and closed her eyes. She remained silent.

Wordlessly, they reached the restaurant. Mario opened her door as Santina smiled. She took his arm as they entered the building. The hostess asked if he had reservations.

Mario replied, "Two for DiLucci."

The girl looked at him with curious expression, "Don't I know you?"

Mario strung her along, "Perhaps, I've been here before."

The hostess smiled as she led them to a table by a window with an excellent view of the bay.

Santina asked as the cocktail waitress approached, "Why did you give a false name?"

Mario wanting to disperse the waitress ordered drinks for them both. The waitress smiled and left. He surveyed the tranquility of the bay. The reflection of lights on the water seemed like stars dancing on the slight movement of the ocean tide. The hero was complaining, "I'm getting a little uncomfortable with all the publicity. I have no privacy. I try to be cordial, but I find it much too difficult to deny my identification."

Santina offered, "I understand but I enjoy the special attention that your publicity generates. Everyone here knows who you are."

When their drinks arrived they toasted each other. Santina toyed with the cherry in her drink as she said, "Your photograph has been on the cover of every major magazine. Using the name DiLucci is useless."

Mario ate the olive and smiled curiously as Santina ignored the cherry. He thought: it may be her way to signify that she is still a cherry.

Pico

"You may be right. This is a classy establishment. The people here respect privacy. Did you know Joe Di Maggio dines in Toots Shore's restaurant and the patrons respect his privacy."

Santina replied, "He's a doll."

Mario added, "The Yankee Clipper has class and most people respect class."

They ordered lobster and, unknown to Santina, he ran his hand along the waitress's leg. She giggled and left, somewhat dazed. Mario took another drink from his glass and gazed absently out the window.

"A penny for your thoughts?" she asked.

He gave her a look that conveyed lusts.

She understood but quickly veered to another subject, "Did you really beat up Johnny Catalano during that fight on the bridge?"

"I've been asked that question many times. How could that possibly interest you?"

"Johnny is my father's first cousin, but we are not on speaking terms with him."

"Why not? Blood is blood."

"Come on, Mario, it should be self-explanatory. Catalano is a dangerous man."

Mario joked, "Don't knock him, he and I were once business associates."

"I hate that man, relative or no relative. He is mean and evil."

After dinner drinks, Mario ordered cognac and sat back and lit a cigarette.

Santana said, "Catalano isn't human. He treats a bullet wound as if he has a cold. The idiot takes an aspirin and goes about his business. Over the years, he has been shot at least seven times."

Mario nearly gagged on his drink, "That means that he has these bullets floating around in his body."

"Yes, that's a fact. The day that someone kills him and that day is inevitable, an autopsy will confirm that."

Mario injected, "He would have been unstoppable in war."

"I doubt it, the pay would not be sufficient for him to risk his life, notwithstanding the fact that his country wouldn't touch him with a ten-foot pole. He also treats knife wounds the same way, like a common cold. The demon has never gone to a doctor."

"How do you know all about the Cat's adventures?"

Santina's Love For A Legend

"I told you, he's my father's cousin. Besides his life is common knowledge."

"Santina, you're funny. The way you talk, everybody is afraid of him."

"Yes, everyone except you, my father and the Padrino."

"Your father doesn't look the type. Next to Johnny, he is a very short man."

"Yes, I agree but he has the courage of a lion. Last year, Cats came into his restaurant and started bothering some customers. He broke a sailor's arm simply because he came from the south."

"Your father should have shot him."

"He didn't have a gun so he scared him with a butcher's knife instead. Catalano laughed and slapped my father in the face. Papa then thrust a knife into his stomach."

Mario wasn't interested with Johnny Catalano but the outcome excited him, "What happened next?"

"Papa ran upstairs as his stupid cousin pulled the knife out and threw it on the floor. He then ran after him and Papa leaped out the second floor window and broke his leg. We haven't seen Johnny since. Grandmother had a talk with him, she's the only person that he listens to."

She agreed, "I'm sorry I brought this up. It is not ladylike."

Mario signaled the waitress to bring the check as Santina said, "You never replied to my question about the bridge fight."

"I beat the living crap out of him. Does that make you happy?"

She smiled, "Yes indeed, that makes me happy."

The manager came to the table, "Mr. Calcagni, I'm honored. Dinner is on the house. I would, however, appreciate it if you autograph the check." Mario signed the check and offered to pay but to no avail. He did leave a ten dollar tip. As they were leaving, Mario received a rousing round of applause.

The drive back was blissful. He sang to the music in the radio as Santina remarked, "They all gave you privacy and acknowledged you as you were leaving."

"I don't really mind that but I still have a problem with all the attention."

It was one in the morning when they arrived at her house. It had a secluded driveway.

The convertible top was now up and he took her in his arms and kissed her. She offered feeble resistance, even revealed spurts

Pico

of passion. Mario knew that she was in love with him but dismissed it as infatuation.

He had had many woman in the past but this girl was intriguing. She was, however a virgin and for the first time in his life, chivalry prevailed. She was willing to accept some advances but nonetheless hoped Mario would not take advantage of her.

Mario reached over and opened the door, "Santina, you are an exceptional young lady. I enjoyed the evening and would love to see you again." He seemed sincere and she was convinced that he meant it.

She kissed him again and skipped gaily to the front door while humming a tune. Santina waved and blew him a kiss as Mario drove away. He had a good feeling about himself. She didn't deserve the five F's.

27

CIVILIAN TRIBULATIONS

The next morning, his mother told him that Patrick had been calling for him all night. Mario called him back and Patrick told him the Mayor had dedicated the Fourth of July as "Mario Calcagni Day," only three days away. Mario thought to himself: here we go again. He sat on the couch as his mother brought him a cup of coffee. She joined him and asked what was going to happen next.

"First, I would like to meet Angelo Micheletti and his family. Then, I am off on a war bond tour with the General. When that's over, I expect to help Attorney General Brown in his bid for the governorship. I should be discharged by then and plan to open a classy restaurant with Benny. Those are my plans for the near future."

"The war is nearly over and I've done enough. They wouldn't allow it even if I offered to go back. Now, you take that general, he wants to return."

Carmela said, "That's understandable, it's his career."

"I suppose you're right, Mom. I thank God that you and Pop are here in America where its safe. It was real bad in Italy. The people were starving. They eat dandelions and wild broccoli. I can't describe what they endured. The old people starved to death."

Carmela asked, "Did you meet some nice Italian girls there?"

"They are all nice, but I didn't have time to think about it. My mind was constantly on survival and the survival of others. I met some brave people, especially in Pico. My uncles, the Michelettis and most noteworthy, a girl named Angelina. She is Angelo's' daughter. A girl of probably fourteen years old but looks about ten."

Pico

"Oh, the poor thing," Carmela said.

"There is something about that girl that I haven't been able to understand. She has the mind of a grown woman. Her personality is dynamic and believe me, she hasn't been abandoned. When I left Pico, I could hear her call echoing in the mountains. I had a strange feeling."

"Mario, my son, I believe that you are in love."

"Mom, don't be absurd. Angelina is only a little girl. We have nothing in common."

"Perhaps you have more things in common than you think," Carmela replied.

Mario shook his head, "I'll tell you one thing, I have never encountered a stronger-minded female."

Rocco came in and yelled out, "Where is everybody?" Without waiting for a reply, he took a jar and went into the cellar to fill it with wine."

I don't know what that man would do if they took the wine away." Carmela said.

"Mom, let him enjoy himself, after all its medicine for him. It can't be all that bad. The church, from the priest to the Pope drink it every day and the Bible mentions it favorably."

"Don't talk that way about your father. He is looking for any excuse he can to keep drinking."

Rocco came up from the cellar and poured himself a large glass as he joined them in the parlor. "When do you want to visit the Michelettis?"

"Let's go tomorrow."

"Good, I'll call Benny to get us some gas stamps."

"Pop, Benny doesn't sell stamps anymore."

"I thought today was his last day," Rocco applied.

"Yes, Pop, today is his last day with that deal."

"What happened to him, did he get caught?"

"No. He and I are going to open a restaurant and I don't want my future partner involved with illegal activity anymore."

"That's understandable, who is taking over his business?"

"He said he is turning it over to Emilio."

"Good choice. Emilio will do a good job. If he gets into trouble, it won't be much of a loss. It's right up his alley. He's crooked anyway."

"How is Hogan behaving?" Mario asked.

Civilian Tribulations

"Hogan is Hogan, He will never change. His poor, naive wife back in Italy thinks he's an angel." Mario didn't want to broach the subject but reasoned that starvation would claim her life.

The next day, Rocco called Angelo and invited himself to his home. That morning, Mario drove Benny's borrowed Cadillac. His passengers included his parents along with Hogan. The trip took two hours in a sweltering heat.

Angelo had prepared a dinner of roasted rabbit with potatoes. His married daughter cooked the pasta. They all lived together along with his son-in-law and their year old son. Angelo had a head start with his drinking and was in high spirits when the paisanos arrived. The men knew each other in Italy and exchanged salutations. His two sons were in the service, both in the Pacific theater. Angelo didn't read the newspapers but he had heard of Mario's exploits from his daughter. He told Mario, "I fought the Austrians in World War I. I have a metal plate in my head."

Mario smiled, "That was a hard-fought conflict."

The old man asked, "Did you get a chance to see my wife and the girls?"

"As a matter of fact, I did and they are fine. Things were a little rough for a while but that is all behind them now."

"I wanted to bring them to America, but my wife is thick-headed. She insists that she had to stay and take care of her parents." Everyone knew the real story about Angelo and Anna.

"That is no longer a problem, her parents are dead," Mario replied.

His daughter began to cry, "Poor Tadone and grandma." She left the room.

"I didn't intend to bring bad news, but these things happen in war," Mario said.

Angelo sighed, "Now maybe they will come to America."

"I'm sure that they will."

They all sat down and ate dinner. Daughter Nina regained her composure. They both asked Mario many questions. He only supplied the good information, and spared them from further anxieties. Mario complimented Angelo on his farm and family. He talked lengthily about Anna, Angelina and Gabriella. Again, the bad times were omitted when they spoke of Pico.

Mario never mentioned the Moroccan Goumiers. He reasoned: What they don't know won't hurt them.

Pico

The Picanos talked about the old times. It was difficult to follow who they were talking about because every family had a nickname. They gossiped and drank wine. It was late afternoon when the visitors decided to leave. Rocco invited them to his house on a future visit and they agreed.

It took a half-hour to finally break away. Mario headed south towards Boston. They were quiet for awhile until Rocco said, "Mario, you didn't tell them much about what really happened in Pico."

"Pop, I didn't have the heart. Besides, they are all fine now. I'll tell you one thing, his wife will never come to America and neither will the girls."

"Why not?"

"I just have a feeling that they will stay in Italy."

"I think you are holding something back."

Carmela poked Rocco with her elbow. Mario didn't answer.

• • • • •

July 4 was a big day for the state. Along with being Independence Day, it was proclaimed "Mario Calcagni Day." What could be more fitting? The war hero being paid homage on the Fourth of July. It was a double celebration.

The evening fireworks were spectacular.

The U.S. Army represented Mario with a drill team that was the highlight of the festivities. Politicians gave speeches in various locations. Mario was showered with flowers.

Ex-Commander Patrick Morgan wore his uniform decorated with ribbons and he never left Mario's side. Attorney General Brown announced that he was running for the democratic governor's nomination and Mario declared that he would help him with his campaign. It was definitely the last day that Mario would be on public display before his home crowd.

The next weekend he met Gen. Biggs in Boston and they headed off on a war bond tour by train that traveled the country.

A gala welcome was repeated at every stop. The troupe was designated as the first team and they sold more bonds than all others combined. The whirlwind tour took over two months. Upon completion of their assignment, Gen. Biggs received his new orders along with a second star.

The general called Mario to his hotel suite. He was smoking a large cigar and had an open bottle of his favorite cognac.

Civilian Tribulations

He was exuberant. "I'm going back to Europe. I got my orders and a new division. What do you say, do you want to come with me?"

Mario laughed, "Thanks, but I've had enough."

"I'll get you a field commission as a Lieutenant. You'll be on my staff. "

"General, if I had to go back, you know that it would be with you. But I've lost my taste for it."

The general poured them both a drink. He said, "Mario, my boy, I was only kidding. From the President and down the line, they would never allow you back into action. By the way, I owe you again. You talked to the President about the other star. I got the promotion and consequently a division command. It's what I always wanted."

"Major General, that's not bad at all," Mario replied.

"Thanks to you, first you save my life and then you get me promoted. Let me tell you, it's important because once the war is over promotions are usually dead."

"Don't worry, General, there will be another war popping up somewhere and you will attain the rank of a full general."

They both sat back in the large luxurious chairs. Biggs asked Mario, "Can I do anything for you?"

"You did enough already. That endorsement deal with the cereal company will earn me a lot of money."

"That was nothing, my family owns a sizable stock in the company."

"You can do me one other favor."

"After the war ends and if you have the opportunity, look up my family in Pico."

"I promise you that I will try my best."

"Please see to it that they and my friends have ample food and proper shelter. That is all I ask."

Both men rose. It was time to say goodbye. Mario gave him a gallant salute and the general finally said, "I owe you my life and all you ask is that I look up your relatives. I want to do something for you and you think of others. If a two-time recipient of the Medal of Honor didn't ask the President, I would never have gotten my division. He could not refuse your request."

They embraced as Mario said, "Be my friend forever and don't get too close to the line of fire, I won't be there to save your ass."

Pico

The general stepped back, "You have my friendship forever."
Mario was misty-eyed as he left.

He was now officially discharged and could now concentrate on his civilian enterprise. Mario telephoned Benny to meet him at the railroad station.

The train ride relaxed him. He thought of Italy and knew that he would go back someday. Something else tugged at his mind. He had it narrowed down to Pico. There was a good feeling that eluded explanation. He slumbered in his seat, realizing that in good time, whatever it was would surface distinctly.

The train pulled into town the next morning and Benny stood waiting. They hadn't seen each other in two months but had been in contact by telephone constantly.

The first thing Mario asked, "Are you staying out of trouble?"

"Of course. Nobody fucks with me because they know that you're my partner."

"Don't rely too much on that because it will wear off someday."

"I doubt it, but for now you're as popular as ever."

"The war is our best advertisement, but when it's over, we lose at least fifty-percent of the popularity."

They drove to Little Italy on the hill. Benny said proudly, "Wait until you see the progress we're making."

Mario interrupted, "I need to meet with Patrick in a few hours."

"No problem. This won't take long."

Benny sped along Calcagni Avenue. The street had green and red lines in lieu of the normal yellow stripes.

"The name of this street alone will keep your popularity high." Benny was whistling a tune as Mario looked at him and laughed. He turned into the parking lot and saw the place bustling with activity.

Mario was impressed. He didn't expect to see the work so far along. It was only two months into the partnership and the construction would be completed by February as scheduled.

As they walked into the building Mario asked, "How the hell did you accomplish so much in so short a time?"

Benny guided his partner forward by the elbow. "It's easy, all you need to do is kick ass once in a while."

Johnny Catalano was standing nearby on a pile of dirt laughing, "All Benny has to do is mention your name and everything falls conveniently into place."

Civilian Tribulations

Benny didn't reply. He felt superior to Catalano and ignored him. Mario didn't worry about Johnny because Cats wanted to be on good terms with the partners. Benny pulled Mario over to a corner, "You know, the Don is here every morning. He has coffee in hand and chats with the workers."

"Is he bothering anyone?" Mario asked.

"No, in fact his presence makes these guys move a little faster. The Don only jokes with everyone, great guy."

Mario replied, "I always liked the Don. He will be good for business. Those guys never bother civilians and they are popular."

The old structure was now taking shape. The entry led into a fairly large waiting area which had a big window that would soon display a large working pasta machine. The opposite side had a similar size window which displayed the kneading and hand-tossing of the pizza rounds.

The dining room section was being prepared for a layer of cement. The final floor would be finally topped with Italian tile. The area was designed to accommodate seating for over two hundred. The kitchen built around the existing ovens were unique because they were fired up from the rear.

Benny had a big chart on the wall that showed progress of the project. He would color blocks as that phase of work was completed.

Mario asked Benny, "How about the cellar? Did they pour the cement floor?"

"Yes, that's one of the first phases."

"The Don was probably happy because the bodies buried there are safely entombed."

"You're nuts, there's nobody down there."

"Any luck with a pasta machine?" Mario asked.

"I found one in New York that needs some work. It will do until we can someday, get a new machine from Italy."

"That won't happen until the war is over," Mario said.

"I have an option on the equipment. I have a great mechanic on hand and he tells me not to worry about missing parts. We will make our own," Benny replied.

"Sounds good, Benny, I'm definitely impressed."

"One other thing, this place is going to have piped in music, but I also plan on live singers."

Mario looked at his watch, "It's getting late, let's go."

Pico

They got in the car and Benny offered, "Too bad you don't have the time, I can get you massaged and laid."

"Sounds inviting, but there is lots of time for that later."

"I bet you haven't had a piece of ass since the last time I took you there a few months ago."

"Benny, I got laid more on this tour that even you can't imagine."

"No shit. Why the hell didn't you invite me along?"

"Go get yourself a Medal of Honor somewhere and you'll be invited on a bond drive."

"Just like that. If I didn't have a wife and kids, I would have been in the service."

"Benny, stick to the restaurant business and if you don't slow down chasing every broad that crosses your path, I'm going to straighten your ass out."

"How about you, the champ of the cock hounds?"

"Benny, I'm not married."

Patrick was waiting as they drove into the driveway. Mario got out and told Benny that he will call him in the morning. He shook hands with Patrick as they walked into the house.

His mother was waiting and had lunch ready. She fussed that her Mario was home. Mario told Patrick, "Well, I'm out and ready to help the politician."

"That's why you're here, the race is close and he needs your help."

"I'll do all that I can. When and where do we start?"

"If your schedule allows, in a few days."

"Commander, I have no schedule. Let's get rolling, the election is less than a month away."

Mario made the difference. The two-month tour was an excellent experience for him. He mastered the art of speaking before large crowds. The Attorney General planned on utilizing Mario's popularity to draw votes. It was a success. The hero was so adept at endorsing his candidate that what was once considered a close race, became a runaway.

Governor Brown was elected by capturing sixty-five percent of the vote. During the victory party, The Governor-elect called Mario to the podium and raised his hand high. He thanked him and in front of the celebrating crowd declared he would not be standing there if it wasn't for Mario Calcagni.

Civilian Tribulations

As Christmas approached, Mario began receiving royalty checks from the cereal company. When his parents saw the first check, they thought it was a joke.

"You should be ashamed of yourself, getting all that money without working," Rocco said.

"Why, Pop? A few more of these and I'll buy you and Mom a small farm."

"That's okay with me, providing you keep it in your name and live with us."

"Whatever you say, Pop."

"Mario, when are you going to find a nice girl and get married?" Carmela asked.

"There's lots of time, I'm still young. I just haven't found the right one yet."

Carmela advised, "There are a lot of girls here in the neighborhood who adore you. We know their families and you grew up with most of them."

Rocco laughed, "You can have your pick. Everybody, and I mean everybody, is constantly asking about you."

Over the next few months, Mario confided to his father about his wartime experiences. There were instances he held back but ultimately revealed most. The information he held back specifically was the horrible incident about his mother. Whenever Rocco asked about her, he replied that she was doing well for her age. He related the experiences encountered by the Micheletti family but made the father promise he would never repeat that Anna was raped by the Goumiers. They agreed that Angelo should never hear it from them.

Christmas Eve arrived and the family held their traditional festivities. The restaurant was nearing completion and everything was going well. Mario kept busy with his social life. He dated many girls but somehow couldn't get serious. He purchased a farm in the northern part of the state. It was forty-two acres with a large pond and the farmhouse had eight rooms with two fireplaces. Some twenty acres were clear rolling hills. There was also a large red barn that required some repair. The Calcagnis moved to the farm in January. It was twelve miles away from Mario's restaurant.

The restaurant opened in February to a gala celebration. The governor was there for the inauguration. The place was an instant success from day one. It became a local attraction. It was predicted to be a tourist destination as soon as gas rationing was over.

Benny was happy as the local money began rolling in. Mario's photographs graced the walls of the building. People from all walks of life came to meet the local war hero. The food was fantastic. Mario made token appearances. He was trying to get out of the limelight, but fame trailed him.

The Don was elated and had a private table set aside for him. He held his important meetings there. Patrick warned Mario that some people were talking about his visible association with the Don. Mario asked, somewhat perturbed, "What people?"

"The people that matter, the individuals that watch society and obey its laws."

"Patrick, let me show you a little comparison. You know Emilio, the Bookie. He takes bets and he is breaking the law, am I right or wrong?"

"Yes, he is breaking the law because he doesn't pay taxes."

"So, Emilio is considered an underworld figure because he takes illegal bets and doesn't pay taxes. Now, take the pillars of society that operate the race tracks all over the country. They dress in Madison Avenue suits and their wives dress like it's Sunday every day. Emilio is a low-life and the others are pillars of society. I ask you, what is the difference between the two life-styles in regards to gambling or to be more specific, exploiting the poor?"

"What's your point?" Patrick replied.

Mario helped him with an answer, "Are you going to tell me that paying tax is the difference?"

"It looks that way. One must carry his own load and pay taxes."

"I agree but that is not the answer. If Emilio pays a fine for booking and he goes to jail, it's because he admits that he is booking bets. The answer is, the so-called pillars of society, the elite who are associated with the power brokers that control society. I say that all the Emilios in this world along with all the pillars of society that own race tracks are all bums. They both do the same thing, they exploit people. In most cases, they remove bread from the table of poor families. The main difference is, the elite have the influence to obtain a legal stamp of approval to exploit the populace. Now take unfortunate Emilio, he doesn't have any legal influence whatsoever but also exploits the public. One segment is a pillar of community and the other is a felon."

Civilian Tribulations

"Mario, I don't know that's a fair comparison."

"It's a known fact that most politicians are crooked. Do I bar them from my establishment because they are thieves? They travel the same highways that the Don travels, only, they use a different vehicle," Mario said.

The commander became uneasy, "You mean to imply that politicians and the Don are all the same?"

"Yes, and if I throw the Don out, I will also throw the politicians out!"

"I think the world of you, but I must say, I don't agree with your approach," Patrick said.

"Is that a fact? It wasn't too many years ago in Boston when a help wanted sign was on display that was accompanied with the initials, I.N.N.A. Patrick, you are older than me by many years, are those letters familiar to you?"

"Yes, it means Irish Need Not Apply."

"So, I ask you, what in the hell is coming off? Has our new Irish governor become a puppet to the power movers? He won without owing anyone anything and now you come around with all this bullshit. Tell my friend, the governor, the Don stays and if he doesn't approve, he can go fuck a horse."

"What has gotten into you, Mario?"

"Nothing, I haven't done anything wrong and I don't intend to do anything wrong. The Don comes in and sits at a private table that he pays for. He doesn't cause any problems and I don't associate with him. The man is a popular figure and is treated the same as any person who is recognized by the public. He is good for business, customers see him and are happy to shake his hand. The man is a crowd drawer."

Patrick took a deep sigh, "The governor conveyed to me that he won't come here as long as the Don's presence is known here."

"The governor is justified in his decision because he is one of the few politicians who is not on the take. Tell him that I understand."

Patrick left with a sad disposition. It hurt Mario to talk that way to his friend. He couldn't comprehend how a person could hold the highest office in the state and yet allow people to control him. He vowed: They can control him but they will never control me. Let them try.

Pico

Benny couldn't believe the success of the restaurant. They grossed over thirty thousand dollars in February. Mario explained, "That's nothing. Next month, we'll have a performer singing Italian and opera songs." A piano was moved into the dinning room to complement the guitar-playing singer. Within a few months, the waiting time for guests to be seated was close to two hours on weekends.

The war ended in Europe and Gen. Biggs and his division performed gallantly. He was near the Austrian border when hostilities ceased. His division was designated to remain in Italy during a brief transition period.

28

GENERAL BIGGS AND PICO

It was mid-summer of 1945 when Gen. Biggs had the opportunity to visit Pico. More than a year had passed since the frontlines crossed the area. The general arrived early on a Sunday morning. Most of the town people were attending Mass. He was accompanied by his staff and an elite squad of soldiers. Mario was informed about his planned visit. The general could have easily assigned the trip to a subordinate, but he had promised Mario and was also looking forward to meeting the rest of his family.

The village was quiet except for the pealing church bells. He ordered his driver, "That's where the action is, pick out the biggest church."

The three-vehicle caravan parked in the square. The lead-car bearing Gen. Biggs displayed a red flag superimposed with two stars. He got out of the car and leaned against a fender with folded arms. One of his staff members asked, "Sir, is everything okay?"

"Yes, this town is so quiet. There was a time when all hell broke loose here and I don't mean war only. This was a German supply area and the front lines moved back and forth. Most of the buildings were damaged and some were completely destroyed. Ironically, the church that is facing us down yonder was untouched."

The major lit a cigarette and replied, "I guess all the mountain towns in this area suffered the same fate."

"Not all, Major, only the ones that were the objectives of the Goumiers."

Pico

"Aren't they one of the French divisions that exploited the Cassino breakthrough?"

"They were indeed and I give them much credit for creating the opportunity for our Fifth Army. The atrocities they committed, however, in all the areas they occupied do not adhere to any standards of warfare." Gen. Biggs unfolded a map and looked to a road leading westward out of the square.

He pointed, "That is the road that leads to Itri and joins Route 6. My friend took that road in reverse order and came here at the height of the battle."

"Was he alone?"

"Yes, he was. His name was Mario Calcagni. We are going to check on his family and a few of his close friends here."

"I was wondering why we drove here."

"Well, now you know."

The Mass was over and people began coming out of the church. At first they were alarmed at the sight of the soldiers but quickly became at ease when it was apparent they were Americans. Their mood became joyous as the general approached and asked an important-looking, well-dressed gentleman regarding the whereabouts of the Calcagnis. He replied that he didn't know them. The general then asked a woman who was walking with a child. She smiled and pointed toward a dirt road that led northward up the mountain and explained that the Calcagnis once lived in town but moved in with some neighbors who owned a farm. The general eyed her with appreciation. "How do we find it? Who do we ask for?"

"They live with the Micheletti family."

"How far up the mountain?"

She smiled, "General, would you like me to show you?"

"Yes, Signora, by all means."

She called a friend who took her child as, she, the Signora Fritatta got in the general's car. La Signora pointed up the curving road as the other two vehicles followed.

Few vehicles ever used the rough lane. It was primarily for mules, donkeys and pedestrians.

The major commented, "Perhaps we should have walked, sir, I don't think this car is going to make it."

The signora explained, between bumps and laughter, "It's around a few more curves."

General Biggs and Pico

Giovanni and Luigi were sitting at a small table near a fig tree sharing a bottle of wine when they heard the noise of vehicles approaching. Italian peasants never worked on Sundays. Giovanni asked his younger brother, "What do you make of all that noise?"

"I don't know, maybe the war is starting over again."

The caravan came into view as the commotion brought all the occupants forth. They halted on the edge of a rolling hill. Gen. Biggs got out and called, "Are you the Calcagnis?"

Signora Fritatta revealed part of a beautiful leg as she said with a smile, "Yes, they are the Calcagni brothers." The general couldn't help but notice.

Giovanni spotted the General's identification on the flag and told Luigi, "What the hell are these Americans doing here, especially a general."

Giovanni faced the soldiers. "We are the Calcagnis ."

The major ordered the squad out of the other vehicles and told them to relax.

"I'm General Biggs and I am a close friend of your nephew, Mario."

The Italians were puzzled and wondered how Mario could have friends in such high places. The general noticed their confusion and roared with laughter, "I was a colonel when your illustrious Mario saved my life in the mountains near Monte Cassino."

"Yes, he told us something about that but Mario didn't talk much about war."

Biggs sat down as Giovanni asked for more wine glasses. The German came into view. During the middle of the conversation, the general called out, "Hello, Horst, how are you doing?"

Gabriella grabbed his arm, fearing her husband was going to be taken away. The general smiled at her and said, "You've got to be Gabriella."

She relaxed and returned the smile. He took a sip of wine and pointed to Anita, "You must be Giovanni's wife and you are the lovely Anna."

He looked at Angelina, "You, I save for last, the most famous girl in the entire Liri area."

Angelina didn't change expression, "Who says so?" she asked.

"I'll give you one guess and I must say, you've sprouted into a beautiful young lady."

Pico

Angelina took a liking to the General but didn't change expression. "You didn't answer my question."

Gen. Biggs replied with an assuring smile, "Who else but Mario Calcagni?"

Anna asked pleasantly, "How do you know so much about us?"

"Mario is my best friend. I was ordered back to the States to accompany the hero. I've been in more parades, honoring Mario to last me a lifetime. We spent two months on tour together and he described and talked about all of you." He laughed, "After the two months, I know you better than my own family."

"How is our brother and his wife?" Giovanni asked.

"Rocco and Carmela are fine. Your brother certainly loves his wife and Carmela, well I can still imagine the embarrassment she displayed when the girls decorated Mario with flowers. He passed them on to her." Angelina displayed concern.

The General sat back asking, "Horst, are you happy with your decision?"

The big German embraced Gabriella and replied happily, "Of course, and I want to thank you again."

"Your thanks are appreciated but your early release is attributed to Mario." Horst smiled with gratitude as Gabriella said, "May God always bless Mario Calcagni. Our prayers are eternally with him."

The General said, "Let's get down to business. How are you all getting along?"

"We're getting by. The fields are planted and we should be okay," Giovanni replied.

"Are you getting any of the provisions being distributed by the United States army?"

"What provision? The power people in town get the first selection and leave nothing for the rest of us," Anna said.

Biggs looked with raised eyebrows, "Are you serious? Who are these bastards?"

Signora Fritatta volunteered, "I'll be happy to point them out. In fact, the well-dressed man that you were talking to is one of the group."

He raised a glass to the Signora and replied, "Thank you." The General faced the group and said, "I have a surprise for you. I promised Mario that I would take care of your needs." He called the sergeant and ordered him to unload the provisions from one of the trucks. He ordered the Italians, "Just tell my boys where to put these provisions. There is no spoilage in any of the goods. You also have blankets and pillows. I could have sent this shipment but I wanted to meet the family and friends." The soldiers soon unloaded enough goods to open a general store.

"What are we going to do with all these things?" Anito said happily.

"Whatever you don't want, give it to friends."

"I don't know how to thank you," Giovanni said.

"It's not necessary. Do you mind if we camp here until morning?"

"Stay as long as you want, we can all have dinner together," Anna replied.

"Why, thank you, I accept the invitation." Gen. Biggs made himself at home. He was truly comfortable with the humble mountain peasants.

The major came forward with two bottles of excellent cognac as Biggs introduced his officer as Major Narducci.

"Oh, an Italian-American," Luigi commented.

"I'm a Sicilian," he answered.

The General ordered the Major to fill their glasses with the prized cognac. He explained, "Mario and I shared many a bottle of this fine French brandy. I confiscated a few cases from a German headquarters that we overran."

"We moved in on them so fast that they left their personal belongings behind," the Major added.

Gen. Biggs held a glass at arms length for inspection and announced, "I feel honored to share this with you." Signora Fritatta flashed an inviting smile that the General caught and returned. He asked Giovanni with a whisper, "Is Signora Fritatta married?"

"She was, but her husband died fighting on the Russian front. She has a small child and like the rest of us, having a tough time." The General whistled and quietly confided in the brothers, "My God, she is a lot of woman."

Luigi added, with a sly smile, "General, I think she has an eye on you!"

He broke into a smile, "God damn it, I get the same impression." The cognac was beginning to effect him and he signaled for more.

Pico

Major Narducci did the honors although he became concerned that his General was becoming intoxicated.

It was mid-afternoon when Luigi began playing the accordion. Angelina and Gabriella sang to the familiar music. The Americans were impressed with the singing voices of the sisters. Neighbors heard the music that resounded throughout the mountains. They had not heard the girls for many years and were drawn to the Micheletti farm. Anna invited them to stay and what started out as a family get-together became a festival.

Angelina walked away sometime later. Luigi noticed her sitting alone under a tree and joined her. He asked, "What is wrong Angelina, aren't you happy?"

"Yes, I am happy on this day but I'm also sad."

"Why are you sad? Is it because of Mario?"

She smiled, "Mario is Mario, I think of him often. The big party we're having is a delightful event, and it reminds me of the wine festivals that were initiated by Tadone. Of all the people who died during this war, I miss my Tadone."

Luigi placed a fatherly arm around her as she began sobbing. "It's okay, Angelina, cry it out."

"Luigi, Tadone heard me and answered me by calling out my name. I heard his powerful loving voice. I called back, he was on the next mountain." Angelina cried. "Then I heard three shots. I screamed out his name but there was no answer."

Luigi comforted her, "I never heard this story before."

"The Germans shot him, they killed my grandfather. It was all my fault. If I didn't find him, he would have been here today. That's why I'm crying."

"Angelina, it's not your fault, very few returned from that forced labor. Most died from starvation. Your Tadone would never have made it back."

"Do you really think so?"

"I know so. Come back to the party, this day is an important change for the future. Enjoy it."

The General began dancing with Signora Fritatta. This signaled that the American soldiers could dance with the girls. Darkness descended as the soldiers built a campfire. Singing and dancing continued into the night. The general had the complete attention of La Signora Fritatta. It wasn't long before she led him into the nearby

General Biggs and Pico

hills. Anna and Anita giggled. They shared the comment that they thought the tradition was reserved for young lovers only. The major was concerned but he certainly couldn't join them.

Later in the evening, most of the neighbors left and the rest of the group settled down around the fire. Even the nightingales joined the singing.

Gabriella began singing "Lili Marlene" in Italian. Angelina followed and the rendition became a duet. When the song was near ending, Horst began singing the song in German. Gabriella joined in as she knew the German version. Luigi played the accordion and the Americans gave their own version. Major Narducci led the squad as they sang the song in English. They all took credit for the music until Horst convinced them that the original was German.

The enlisted men retired except for one posted on guard duty. The General and the major sat around the fire with the family. Luigi added wood to the fire as the conversation centered on Mario. Giovanni wanted to know what Mario was presently doing. General Biggs offered, "I haven't seen him in nearly a year. I did get a letter from him a few months ago and he co-owns a successful restaurant. The people there named a street after him."

"Does Mario have a lot of girl friends?" Angelina asked.

Biggs replied, "How can they resist him? The girls are all after him, the man has everything going for him, fame, looks and charm. He is also intelligent. He is now making big money. However, a strange thing about Mario. He dates often but he never gets serious with anyone."

"Perhaps the right one hasn't crossed his path," Anna said.

"I'll tell you this, some beautiful girls in America crossed his path when we were on the war bond drive."

Angelina was listening intently as Anna asked, "And what happened?"

"Like I said, he never got serious. Something is blocking the way."

"Maybe he is like his uncle Luigi here who hasn't married yet," Giovanni said.

Luigi replied in defense, "I'm only thirty-two years old and the right one hasn't crossed my path."

They all had a good laugh as Anna kidded, "I wonder how many illegitimate children Luigi has roaming these mountains."

Pico

"Stop picking on my brother-in-law," Anita said.

Major Narducci passed around cigars. He lit the general's cigar. Biggs laid back and sighed, "This is the life." He enjoyed the company and he looked at La Signora Fritatta and said, "I think we will stay here a few more days. We've got to straighten out the mess in town." Signora Fritatta smiled as the General asked, "What are you smiling about?"

"Nothing," she replied.

General Biggs asked her, "What about your child?"

"Don't worry about a thing. That girl that I told you was my friend is also my sister. My child is in good hands. It's very late and I must be getting home, they will be worried."

The general replied playfully, "Why don't you give them a telephone call?"

La Signora Fritatta turned to Anna and Anita, "This general is crazy. Tell him, we don't have a telephone in the entire town."

The General laughed, "I'm only fooling. We can put you up in a tent."

"No thank you, I must go. I will look for you in the morning."

"Are you intending on walking back to town alone?"

"Of course, it's perfectly safe."

He motioned to the major to awaken a soldier and accompany Signora Fritatta to her house. Biggs kissed her good night.

"The general is in love and all in one day," Anna said.

"That is the Italian style," Anita answered.

The major sadly added, "He is married back in America."

"How does he handle this situation?" asked Giovanni.

Major Narducci explained, "She is a good general's wife. She knows what's going on but looks the other way."

General Biggs came back to the campfire as Major Narducci inquired, "How long are we going to be here?"

"It could take two or three days to correct a problem that I didn't know existed," he replied.

The major knew that the task could be cleared up in a few hours. They all laughed including the general. Narducci said jokingly, "Perhaps I should look for a girlfriend for myself."

"What are you waiting for? Get some enjoyment now out of life."

The festive occasion was over. The nightingales now had the night to themselves. Everybody retired after one last final toast.

General Biggs and Pico

Angelina couldn't sleep. She tossed and turned and could not stop thinking about Mario. She was happy to know that he wasn't serious with any of the available American girls. Angelina was saddened by the fact that he was absorbed by success and would probably never return to Italy. She thought, even if he visited, he would probably neither settle here nor marry her. In the girl's mind she imagined: In a few years I will be even more beautiful than Gabriella.

Angelina was assessing herself short, she had already surpassed Gabriella's beauty. Yet, she knew that the attraction would not be beauty. It was only a matter of time when Angelina becomes a lady.

They all arose early the following morning. The soldiers were soon brewing coffee as the women began breakfast with the provisions they received. The Americans left one soldier behind and headed for town. The General ordered, "Leave the car here, we'll take the trucks."

The peasants went to work in the fields and the children followed.

It was eight in the morning when the Americans arrived in town. Signora Fritatta heard the trucks and left her house to meet them. The soldiers came down from the trucks and lined the square. Known as Archefanti, Signora Fritatta pointed out the building that housed the corrupt officials. The general and the major entered the establishment. One person sitting at a desk didn't bother to look up to acknowledge the visitors. The general kicked the chair and the occupant went sprawling to the floor. He began cursing until he recognized the people and began saluting awkwardly.

"Who the hell is in charge here?" General Biggs asked as he brushed the top of the desk with his riding crop.

Two cronies came running out of a rear room, cursing until they saw the American officers and stood in attention.

The general slashed the desk top again for emphasis and ordered, "Don't ever fucking salute me again. I want every ounce of provisions that you have stockpiled moved out into the square immediately."

One of the men pleaded, "General, we have very little and we are rationing it as best we can."

General Biggs removed his forty-five pistol from his holster and pressed it against the Italian's forehead, "Do you understand me now?"

Pico

The man turned white, "Yes, I understand."

"Get a move on."

The major called in six soldiers to help them move the goods. The general went outdoors and requested Signora Fritatta to send someone to the church to ring the bells. He also assigned some youngsters to pass the word in the mountains that provisions would be available to all. The signora observing her new hero thought to herself: Now I know why he is the General.

The bells rang and people began to assemble in the square.

Gen. Biggs turned to Signora Fritatta and told her, "I need some honest people, send someone up the hill to get Giovanni, Luigi and Horst." The provisions were being lined up along the front of the building. The peasants couldn't believe the amount of goods that came out of the building. He placed the three men under arrest and began an immediate investigation.

The Calcagni brothers and Horst arrived as the general explained his plan. He wanted every family to have an equal share of the provisions. The general asked them, "Why is it that you people never received any of these provisions?"

Giovanni replied, "From what we heard, these corrupt idiots were holding back seventy percent of the shipment that they later sold on the black market. They distributed the remaining thirty percent. The peasants in the mountains were last in line and received nearly nothing." The trio was placed in charge of distribution. Anyone who illegally purchased goods did not receive ration. The Americans had no trouble identifying them because there were a large number of townspeople who came forward with information. Major Narducci supervised the operation as the general quietly slipped away with Signora Fritatta. She now became an important lady in the community. The general was a hero, who, in a now short period of time, found a home. La Signora was a widow and now the general's mistress.

The distribution was completed by late afternoon. Everybody was happy except for the prisoners and the civilians who purchased stolen property. Gen. Biggs hadn't decided their fate.

That evening, a committee was formed to organize a celebration in honor of the American general and soldiers.

They planned the festivities for the following day but Major Narducci informed them they had to leave before then. Angelina

General Biggs and Pico

heard the major and decided to talk to the general. She knew where the lovers were. Signora Fritatta would normally sit under a large oak tree that had a panoramic view of Roccasecca and the Liri valley.

La Signora was sitting with her back against the tree. The general's head was comfortably set on her lap. Angelina approached and heard that he was speaking English and she was speaking Italian. From what she could make out, they didn't comprehend each other but somehow seemed to get along.

La Signora spotted her coming up the trail, "Angelina, how nice to see you, come sit with us."

She sat near them and related, "The British advanced along that area on the other side of the river. If only they had been on this side, we could have avoided the Moroccans."

Gen. Biggs sat up and said, "You're right, that area was the dividing line between the British Eighth Army and the American Fifth Army."

Angelina asked, "When are you leaving?"

"I would like to stay here for a few weeks but I have a division up north that I'm responsible for. We should be pulling out of here tomorrow."

The general would have grasped at any legitimate excuse he could, but was at a loss. He wanted to remain a few days longer but couldn't justify it.

Angelina said, "You can't leave tomorrow."

"Why can't I?"

"Because the people want to hold a festival in your honor. They appreciate what you have done."

"That is not reason enough," Gen. Biggs replied.

La Signora spoke with an alluring voice, "Angelina is right and what about me, caro mio?"

The general was searching his mind for a reason as Angelina said, "You must establish an honest government here and that takes a little time. You also must deal with all the corruption."

He asked, "How old are you?"

"I'm over fifteen, why?"

"You look far beyond your years. You've given me the reason to stay a few additional days. Let me ask you young lady, who do you recommend as the new officials of Pico?"

"The Calcagni brothers."

Pico

"Convince me?"

"They are honest. They are well-liked by everyone. They helped Mario against the Moroccans. You can trust them with your life."

"Makes sense, Angelina. Can you do me a favor?"

"If I can."

"Go back to town and tell the people to go ahead with the festival."

Angelina ran down the trail singing her favorite song as the General shook his head in approval, "I can understand why Mario is in love with that girl."

Signora Fritatta gave him a long kiss and asked, "Did Mario tell you that he loved Angelina?"

"No, he never came outright with it."

"Well, how do you know?"

"I can't explain it but somehow I could sense it. Whenever he mentioned her name, and it was often, there was something special in his voice."

"My General, you are a good man." She sobbed, "Even though I have a child, my life has been empty. You have given me the inspiration I needed." She kissed him passionately.

Between kisses he said, "I don't know what it is about you Italian woman. I shall miss you dearly."

He embraced her as they rolled in the grass.

The celebration was a great success. During the festivities, Signora Fritatta called out, "Angelina, come, I must talk to you."

They walked arm in arm while viewing the various fruit and vegetable stands. Angelina commented, "You appear to be very happy, is it the general?"

"Yes, I think that I'm in love but I am also happy for another reason."

"Does it concern me?"

La Signora replied, "As a matter of fact, it does."

Angelina's eyes brightened, "Tell me, what happened?"

"Oh, nothing much except that General Biggs told me Mario is in love with you."

"But Signora, how could the general possibly know?"

"Well, he has observed that Mario speaks often of you without realizing it."

General Biggs and Pico

Angelina became serious and asked, "What do you think?"

"My opinion is more of a fact than myth. Mario Calcagni is in love with Angelina Michelitti." Angelina kissed Signora Fritatta on the cheek and returned happily to her friends.

With the approval, the Calcagni brothers were placed in charge of Pico. The corrupt officials along with their families and belongings were put on the Itri road and were told not to stop. The General explained, "Naples will absorb them."

The following year General Biggs made numerous excursions to the mountain village. The main reason was Signora Fritatta. On other occasions, she visited him in the north. The relationship was becoming serious indeed. The General had thrown caution to the wind.

29

GOODBYE TO A GREAT IRISHMAN

The telephone rang as Carmela was preparing breakfast. She wiped her hands on a towel as she picked up the receiver and said, "Hello." The voice on the other end of the line was sobbing but she recognized it to be Patrick's wife.

"Etta, what's wrong?"

"It's Patrick, he had a massive heart attack."

"Where are you?"

"I'm at the hospital."

"Hang on, I'll get Mario." She placed the telephone down and went into the bedroom to awaken him. "Mario, wake up, you have an important call."

He rolled over and said, "Take a message and I'll call back."

"You don't understand, it's Etta and she is crying."

Mario leaped out of bed and rushed to the telephone, "Etta, what happened?"

"It's Patrick, Mario, he is dying?"

"Oh my, where are you?"

"I'm at the hospital."

"I'll be right there."

He hurried back into the room and began dressing. Suddenly it came to him that, in the confusion, he forgot to ask what hospital. He swore, "Fuck, I don't know what hospital, now I've got to waste time with the yellow pages. Where is the telephone book?"

She handed it to him and said, "Calm down, there is nothing you can do."

He opened the book and called the first hospital. The operator answered and Mario asked if Patrick Morgan was admitted there.

"I am only the operator, I can direct you to another department," she replied.

"Wait, my name is Mario Calcagni. Can you do me a fast favor?"

"Of course, anything."

"Can you find out if Patrick Morgan is there and if he isn't, find out where he's at?"

"Just give me your number and I'll call you right back."

He gave her the number and hung up. Mario hurriedly washed his face and brushed his hair. The phone rang and the operator informed Mario, "He has been admitted to General Hospital."

"Thank you, come in and see me at my restaurant."

"I certainly will."

Mario hung up and was out the door and driving away as his mother cautioned, "Don't drive fast." He drove to the main entrance and parked in a 'No Parking' section.

Ignoring a security guard, he hurried to the intensive care ward and saw a doctor talking with a sobbing Etta. The doctor held her by the elbow and was leading her to a sofa.

Mario came to her and she began, "He's dead, my Patrick is dead."

Mario embraced her, "Etta, I'm so sorry, what can I do?" He tried to wipe her tears with her hankie.

"You'll never know Mario, I loved him so much. He is all that I've ever had. I'm alone, I have no children."

"Etta, you're wrong, you have me. Patrick was my second father. I could never repay him for what he did for me."

"Patrick always said, you owe him nothing." She sobbed, "You were the son that he never had."

"I know. I promise you that I will forever watch over you. I will be the son that you never had." She kissed him on the cheek.

The Governor arrived and offered condolences to his sister.

Mario told him, "I'm taking her home to my house. I don't want her to be alone."

"Thank you. I'll see that the proper arrangements are made and will send someone to get her in a few hours."

Mario embraced the governor. He placed an arm around Etta and led her out to the car.

Pico

It was one of the largest funerals ever in the state. The Navy was represented as pall bearers. Etta was presented with a flag of the United States. The mourning caravan consisted of over one hundred fifty cars.

In the ensuing year, Mario visited Etta at least twice a week. One of her cousins invited her to move in with them but she declined. Etta was a very independent woman and became active in the neighborhood

The old Navy commander was Mario's best friend and whenever she wished, Mario would drive Etta to the cemetery.

He would leave her alone at the grave. From a distance, he could hear her talking. During these periods, it appeared that she was having a normal conversation. From her reaction, Mario could swear that Patrick was answering Etta.

Benny was another story. He had a unique way of occupying other people's time, especially Mario's. His wife, Linda was contemplating a divorce. Every time she neared that goal, she would call Mario and cry on his shoulder. One morning, she called him and asked if he would join her over a cup of coffee. He agreed and they met on the Boston Post Road. He knew Linda since he was a child and always felt bad for her because Benny was a womanizer. They ordered coffee with pie.

"What's wrong, Linda?" Mario asked.

"It's Benny, he hardly ever comes home."

"I know Benny has always been a little wild. Does he give you enough money?"

"That has never been a problem, he takes good care of the family."

"What's he done lately to get you riled up?"

"He knocked-up a nineteen-year-old girl who intends to have his baby!"

"That could be a rumor. Benny gets a lot of credit for things that he doesn't deserve."

"I know for a fact because the girl called me. Mario, I was open-minded about the whole thing."

"You were. Why?"

"Because I know Benny. I could never satisfy him enough and I have always known his reputation."

Goodbye To A Great Irishman

"What did you tell the girl?"

"I told her to get rid of the baby and do you know how she answered? That *putana* had the nerve to tell me that she is a good Catholic." Mario nearly choked on his coffee and tried to hold back the laughter. Linda couldn't help but laugh herself.

"Did you confront Benny about your conversation with the *putana*?"

"Of course I did. He said he never heard of her and that malicious people are always trying to get him into trouble."

"Linda, you've got to see them with your own eyes before you fly off the handle."

"I saw the son of a bitch on two different occasions and his *putana* was wearing a skirt where the hemline was up to her ass."

Mario shook his head and covered his mouth with a napkin to hide his hilarity. Linda continued, "They look so stupid. She is about six inches taller than he and when she wears high heels she towers over him by nearly a foot. Can you imagine that bastard walking arm in arm with that *putana*? From the rear, they look like Mutt and Jeff. You can hardly see him. The only thing that stands out is her nearly bare ass."

Mario laughed because the picture she painted in his mind was comical. "Did you tell Benny that you saw him with her?"

"Yes. He claims that I need glasses and offered to make an appointment for me to see an eye doctor."

"Linda, what do you want me to do?"

"Talk to him. I don't care if he's out screwing around. I don't need the aggravation of him knocking up *putanas*. Benny has children and it doesn't look good."

"I'll talk to him but I can't promise anything. Benny doesn't listen to anyone."

"Like hell! He may listen to you because he respects you."

Mario and Linda left the diner. He walked her to her car. She kissed him on the cheek and said, "Tell Carmela that I'll visit her soon."

Mario drove to their restaurant. Benny was in the parking lot playing the game La Amorre with an old Italian who beat him quite easily.

Mario waved him into the office. It was winter and appropriate to drink *coffee royal*, a combination of espresso and anisette. Mario

Pico

called a waitress and ordered two. They sat down and Benny stretched out and yawned, "I'm fucking tired, I need a little vacation."

"You're right, that young broad that you're fucking is making an old man out of you."

"Hey, that's the best piece of ass I've ever had, and she gives terrific blow jobs."

"Yeah, and she's also pregnant."

"So what?"

"Benny, you sure are smart. Are you planning another family?"

"Why not? I've got a lot of money."

"What about your wife, are you going to divorce her?"

"No, Linda is a basket case. She doesn't know what's going on."

"Everybody knows you screw around and we always laughed about it but I think you're out of control."

"Mario, you're talking to me, the champ. My dream is to die in bed with three broads around me."

"I advise that you shut your *putana* up because she called your wife and keeps Linda up to date."

"That's bullshit, she doesn't have the telephone number."

"Pay her off and get rid of her."

"How in the hell am I going to do that? The girl is in love with me," Benny replied.

"Don't make me laugh, she's only in love with your money. Benny, I told you many times, if you pay a broad every time you fuck her, you make her into a whore and there should never be any repercussions. You want to play big time by thinking that your charm or whatever you want to call it, got you the action."

Benny became aroused, "I never paid for it. I take great pride in nailing a broad."

"You're full of shit. Those days are gone. I'm telling you right now, "get rid of the girl. Pay her off."

"It may be too late, I'm in too deep with her."

Mario offered, "I'll talk to her. We'll pay her off, everybody has a price."

"Go ahead and good luck."

"From now on, Benny Boy, every time you fuck a *putana* or anyone else, pay her. Your wife won't find out and even if she does, she'll look the other way."

286

Goodbye To A Great Irishman

Benny saw the look in Mario's eyes and realized that he was serious. He rose up to leave but before reaching the door, he turned and said, "Okay, talk to her."

Mario smiled, "Benny, we have a great thing going here, let's not jeopardize it."

"You're right kid, thanks." Benny knew the truth but would not face reality. In the future, he paid the ladies every time.

Mario gave the pregnant opportunist a gift of five thousand dollars and ordered her out of town. She complied and was never seen again. When Linda called Mario to thank him, he told her that he knew nothing about it. Benny was his good friend and he wouldn't embarrass him.

30

CATALANO'S REVENGE BACKFIRES

The telephone by Mario's bed rang. He turned on the night lamp and saw it was nearly three in the morning.

Benny's wife, Linda, was on the line. "Benny isn't home yet," she complained

"Linda, you know Benny, he is probably shacking-up with a waitress."

"Yes, I know. When he does things like that, he gives me a big line of crap as an excuse for coming home late. He didn't say anything when I last talked to him."

"Did you call the restaurant?"

"Yes, but the line has been busy for a few hours. I think the telephone must be off the hook."

"All right, if it makes you feel any better, I'll go check it out."

"Thanks, Mario, I knew I could depend on you."

Mario dressed and drove to the restaurant. The lights were on as he unlocked the front door. He stepped into the building and couldn't believe his eyes. The interior of the establishment appeared to have been hit by a hurricane. Tables were turned over, chairs were busted. The air reeked of alcohol because most of the liquor bottles were smashed. The mirror behind the large bar was shattered. Mario made his way through the debris to the office. The room was in shambles.

He heard a weak moan and saw Benny. His partner's face looked like it passed through a meat grinder. Mario tried to revive him but to no avail. He poured water into a bowl of ice and used a towel to soothe Benny's distorted face. The cold towel awakened him.

Catalano's Revenge Backfires

"Benny, what in the hell happened?"

He stuttered the words, "It was that fucking Johnny Cats. He was looking for you and he said that I was the next best thing. The man is an animal."

Mario poured a shot of cognac and lifted Benny's head, "Drink slowly, it will do you good."

"We all know he's an animal but what the hell is his beef?"

"I don't know, the only thing I could figure out was that you're fucking his cousin and he is defending her honor."

"The bastard is crazy. I've been out with her but I never screwed the girl. She hates him, doesn't even talk to him."

Mario helped his partner into a chair as Benny ached with pain, "My ribs must be busted, I can hardly breathe," he moaned. "The son of a bitch is a raging wild animal. I put a slug into him but I don't think the fucker knows I shot him."

"You should have shot him in the head."

"Mario, his reason is the fight you had with him. He won't rest until he gets even. You know, the bastard is stronger than ever and is determined to kill you."

"Benny, I beat the shit out of him when I was a kid. Do you have the slightest thought that I fear that asino now?"

"Kid, sometimes, I forget who you are but you better start carrying a gun because he will try to nail you."

Mario carried him to the car and drove to the hospital. He called Linda and informed her that her Benny was in a small fight but was okay. She cried, "You're not telling me the truth, I know that he's hurt."

"Linda, I'll have him home in a few hours."

It took nearly two days to restore the interior of the restaurant. Benny was in bad shape and consequently stayed away from extracurricular activities. It was rumored that Johnny Catalano closeted himself in a New Hampshire motel to rest and heal his gunshot wound. The word was out the next morning. The entire community was aware of Johnny Catalano's latest escapade.

The Padrino was furious. Cats was overstepping his boundaries. He was repeatedly warned but paid no heed. The Padrino had sworn proof that Cats was shaking down local merchants for protection money. Johnny had the impression that he was well-entrenched in the organization and that he was untouchable.

Pico

The Padrino, however, was known to extend a long rope before deciding to discipline a subordinate. Once that point is reached, the course of action was cast in cement.

Mario was inquiring for Catalano's whereabouts. All he could visualize in his mind was another Goumier and he was prepared to take care of business. Johnny bit off more than he could hope to chew. Benny knew Mario's determination. He feared that Mario's life was at a crossroad and as events unfold, the hero could end up in jail. Benny decided to speak of his concern to the Padrino.

The mafia chieftain was having lunch at his private table when Benny asked, "Could I talk with you privately?"

"Sure, Benny, sit down." He signaled his bodyguards to allow him privacy.

"How are you doing, young fellow?"

"I'm doing fine, Padrino, but I'm worried about my partner. He's made up his mind to kill Catalano and heading for deep trouble. I know Mario too well, he will do it out in the open."

The Padrino placed an assuring hand over Benny's fist, "Don't worry, your friend who is also my friend, will never get the opportunity. Mario is the best thing that ever happened to all of us. I won't permit him to dirty his hands."

"Padrino, I don't want to hurt your feelings but Mario won't listen to anyone, not even you."

The Padrino deviated and laughed, "I don't expect a man of Mario's character to listen to me. Benny, do you trust me?"

"Of course I do. Everybody trusts you."

"Then don't concern yourself any longer. Johnny Catalano's fate is out of Mario's hands. He doesn't know it but Cats will soon be history."

"Can you tell me more?"

"Benny, Benny, I haven't told you anything. Go about your usual business, everything will turn out fine." The Padrino ended the conversation with a slight movement of his hand. The look in his eyes revealed that Catalano's days were numbered. Benny had the feeling that he no longer needed to be concerned.

Mario was having lunch in the office as his partner entered.

"Hey big guy, what are you eating?" Benny asked.

"My father harvested some ocean snails with this morning's tide and steamed them in a pressure cooker."

"How are they prepared?"

"It's a seafood salad, olive oil, lemon juice, scallions and garlic. What the hell, are you writing a cook book?"

"I love snails but they could be tough on my teeth, they still hurt."

"Sit down and have some. I told you that the pressure cooker softened them."

Benny sat and joined him. "Why are you eating alone here in the office?"

"I've been thinking. Nobody has seen nor heard anything about that fucking Catalano. The son of a bitch has disappeared into thin air."

Benny tried to convince him, "Hasn't it occurred to you that the bullet I put into him could have killed the bastard?"

Mario smirked, "You need a bazooka to permanently get rid of that fucker."

"It was only a thought, but who knows, he could be dead," Benny replied.

Mario drank some wine, "Rest assured, Johnny will emerge and disturb the peace."

"All in all, I hope he is dead."

"I hope that you are wrong," Mario replied.

A few weeks passed and Johnny Catalano finally returned. Apparently, his wound healed and all he needed was a few drinks to start throwing his weight around. Cats entered the Bella Notta bar at ten one night. He began drinking and didn't bother anyone as he quietly occupied a bar stool. The three seats on either side of him were vacant. The music from the juke box entertained the patrons.

A well-dressed man entered the bar and walked to within ten feet of Catalano. Johnny never saw him as the man suddenly raised an automatic weapon and fired at Cats. Every bullet entered the body. The music continued as the customers froze in fear.

The body lurched sideways and finally rested on top of six bar stools. The gunman placed another clip in the weapon and fired again. The inert remains quivered from the impact. Johnny Catalano was dead.

The gunman spat at the body and quietly left.

The music was still playing when the police arrived. They sealed the area and detained the witnesses. There were twenty-one customers along with a bartender on the premises. Not one person could identify the gunman. Some said it was too dark. Others stated they didn't see anything. It was later rumored that a local man made the hit.

The police made a lame effort to investigate. After all, they opined, Johnny Catalano was bad news. Benny was relieved and Mario offered no comments. The autopsy revealed seven previous inflicted bullets in Johnny Catalano's body.

31

BURNING DESIRE

One day in the spring of 1947, Mario was sitting with his father on the patio of their farm. Mario was quiet. They were drinking wine with some goat cheese and bread. The sun was setting and streaking pink and yellow hues through the tall pines that lined the green field. The birds were singing as a flock of ducks flew overhead and made three passes before landing in the pond.

In recent weeks, it had become nearly a ritual for Mario and Rocco to watch the evening sunset.

Mario told his father, "I think you and mom should make a trip to Italy."

Rocco replied. "I'm not ready yet."

Mario continued, "Your mother is getting old and one of these days you're sure to receive some bad news."

Rocco looked at his son and sensed that Mario was holding something back. True enough, Mario was debating whether to tell the full story about his father's mother. Of course, his brothers and everyone in Pico knew about it. Sooner or later, Rocco may hear it from a stranger so he might as well find out from his son. Mario also reasoned that knowing the truth may prompt them to visit her before she died.

Well, Mario thought, tonight might as well be the time to let him know. Mario held his father's hand and told him the entire story. Rocco put his head in his hands and listened without interrupting. When Mario was finished, Rocco rose and walked into the house.

Pico

The next morning Rocco came out to the patio. Mario was having a cup of coffee as he admired the beauty of the rolling hills. The birds were singing nearby and some were wading in a puddle formed by an earlier rain. Rocco sat across from his son as Carmela brought out a coffee pot with more cups. They sat together and enjoyed the view and the serenity of the area.

"What are you thinking about?" Rocco asked.

He looked at his parents and said, "There is something missing on this farm and I finally figured it out."

"We have chickens, a few goats and a dog. I don't want to be babysitting a zoo," Carmela replied.

"What is missing is horses. I've been so busy with the restaurant that I haven't thought about it. There is nothing more graceful than beautiful thoroughbreds running in those fields. What do you say, Pop, do you want to be a horse breeder?"

"I never thought about it but the idea isn't farfetched."

"I miss riding. The barn is in place and requires very little change. All we need is fence to enclose a pasture area," Mario said.

"Who is going to take care of them when I'm not around?" Rocco asked.

"I've thought about that. First, you don't need to work anymore. You can become a gentleman horse breeder. Secondly, this farmhouse is so big, why don't we get Hogan to send for his wife from Italy?"

"Not a bad idea, he finally located the poor soul and wants to send for her," Rocco asked.

Carmela added, "Can you imagine the character of those old women? After all these years, she has waited for his call."

"Hogan is getting old and this place would add years to his life but how are you going to keep him on the farm?"

Mario countered, "I'll find a way."

"Good luck! Between the pool room, the racetrack and your restaurant, Hogan is in his own world."

Mario poured more coffee into his cup and reminded his parents, "A few months ago, I mentioned that I wanted to revisit Italy."

"You said that you didn't know why so I thought you forgot about it," Rocco replied.

"I haven't forgotten. I think about it every day but I still can't figure out the urge that compels me. Perhaps I want to see and

Burning Desire

enjoy the area without the horrors of war. Maybe it's the people, they were all great. It could very well be the serenity of the mountains. Perhaps I want to walk the trails that are now safe. There is also a possibility that I can help them from the ordeals they've encountered over the years."

"You should take a little vacation and get it all out of your system," Rocco advised.

"Pop, we have the money to do whatever we want. Last week, I received a call from a publishing house that wants to do a biography on my war experiences."

"Good idea, did you accept?"

"I've decided to accept, but I haven't told them yet."

"How much are they going to pay you?"

"One hundred thousand dollars plus royalties."

Carmela was speechless.

"They have assigned a writer. All I need to do is sign the contract."

"Mario, what the hell are you going to do with all this money?"

"Get some good horses on this beautiful property. I'll put Hogan on the payroll and get him out of the city," Mario said.

"You can also hire a few farm hands."

"When you mentioned a few horses you really mean quite a few horses." Rocco asked.

"Yes, the first thing to do is get a good bloodline stallion for breeding. The next thing is the proper mares. I think you will both enjoy it yourselves."

"I can just picture your mother on a horse," Rocco bantered.

She playfully slapped her husband on the side of his head with the morning newspaper.

"I talked briefly with the writer. He wants to leave for Italy in a few weeks," Mario said.

Rocco smiled, "It's been three years since you were there. It's time you went back." He then asked, "When are they going to give you the money?"

"As soon as I sign the contract, the advance amount will be in our account on the same day."

"What are you going to do about the restaurant?"

"That is no problem. Benny can be trusted. He only needs a kick in the ass every once in awhile. That's where you come in,

Pico

Pop. Keep an occasional eye on the place while I'm away. Throw a scare into Benny as the situation demands."

"What do you mean, how?"

"Just grab him by the throat but don't hit him. He will get the message."

"I don't know Mario, I don't feel right going in there and giving Benny a hard time."

"Pop, Benny knows that everybody there works for me. He and all the employees there will be advised that you are taking my place. Remember, that business is yours as much as it's mine."

"Mario, parents couldn't ask for a better son. Your Mom and I don't require all the things that you're doing for us."

"Mom, Pop, enjoy life because it's too short."

"Besides horses in the field there is another thing missing," Carmela said.

"What's that, Mom?"

"A daughter-in-law and grandchildren. Mario, people are talking."

"Let them talk all they want. It doesn't bother me. I'm only twenty-four years old. I have met a lot of girls but I am still looking for the right woman."

Carmela forced the issue, "You're going back there and you tell me it's about a book. I believe that's only part of the reason. How long will you be gone?"

"Mama, I'll be there as long as it takes."

"I'm your mother. I carried you in my womb for nine months. I know you Mario and I know there's more to this trip than meets the eye. Is it another girl?"

Rocco looked sharply at Carmela as if to admonish her.

Mario let out a deep sigh and embraced his mother, "To tell you the God's honest truth, I don't know if there is another reason."

"Mario, I know it's another girl so what are you going to do about Santina?"

"Don't be ridiculous! How can you stand there and tell me that I'm involved with something that I have no knowledge of?"

"Mario, you will never get that out of my mind. I know and I can't tell you why or how."

Rocco interjected, "Don't pay any attention to her, she's crazy."

Carmela never mentioned the incident, but she heard Mario talking in his sleep and he seemed to be terrified with something

Burning Desire

about an Angelina. Presently, she demanded, "You never answered the other part of my question. What are you going to do about Santina?"

"There's no problem with her. She's a great kid and if I were ready to marry...," his voice trailed off. "I'll talk to her over dinner."

"She will make a wonderful wife," Carmela said.

"I agree, she has it all but I've known her for only about six months and everybody thinks we're getting married."

Carmela pressed, "When are you going to get married?"

"I told you before, I'm only twenty-four and not ready to settle down."

"Your father and I would appreciate some grandchildren and without a wife, it's not possible."

"She is sure feisty this morning, besides, you don't necessarily need a legal wife to have children," Rocco said.

"Pop's right, Mom."

Carmela smiled and threw a towel at her son. "It's a girl and that is the reason you're going back."

"At least, you can't go wrong with a greenhorn from Italia."

Mario opened the refrigerator and was looking for nothing in particular as Carmela ordered playfully, "Shut that door before all the food spoils."

"It's a hot spring day, perhaps we can cool off the house," he joked.

"How are you getting to Italy?" Rocco inquired.

"The U.S. Army. It's good public relations for them. They are flying me and the writer to Naples. I'll probably just borrow a jeep and drive the same route to Pico that I previously walked."

"Does the writer know?"

"It doesn't matter, he will do whatever I say."

Santina sensed a troubled Mario's as they were enjoying a quiet dinner. Mario was busy with the appetizer of little neck clams on the half shell. He applied hot sauce and lemon which produced a curling effect to the raw shell fish. Mario handed her a half shell and advised, "It's good for you."

She smiled, "So I've heard, but I'm not a man."

"Do you actually believe all that hogwash about its potent sex power?" Mario asked her.

Santina changed the subject, "What's wrong? You've been unusually quiet since you called for me at home."

Pico

"I don't know, I guess its a lot of things catching up on me. All this crap about your cousin Catalano, Patrick Morgan dying. Now, the upcoming documentary on T.V. The book about my life. It's all catching up on me. I'm getting tired of all the attention."

"You've handled it very well."

"No, I really haven't. I conceal everything inside."

Santina sipped her Tom Collins and asked, "What do you propose to do?"

The background music was a Sinatra hit song and Mario said, "There is a paisano that has it made in the shade." He took a drink of cognac. "Do you know why I always order this?"

"No, why?" She was satisfied that he was now normally conversing.

"It's General Bigg's favorite drink. He is a true and trusted friend and I drink in his honor."

"I assume that he will be you best man when you get married," Santina said.

"Yes, that's a fact."

"You haven't answered my question. What do you plan on doing?"

"I'm going back to Italy next week, with a writer."

Santina fidgeted with the cloth napkin, "How long are you going to be there?"

"I don't know exactly, I may stay awhile and call it a vacation. I have family there that I want to enjoy under peaceful conditions."

She looked away as tears formed in her eyes. Woman's intuition was penetrating her mind as she tried her best to cover her concern. He couldn't help but notice her reaction and tried to veer to a new direction. "The swordfish should be here soon." He looked around and wished the waitress would appear.

Santina pressed, "We have been going together for over six months and you never told me you loved me."

He thought for a moment and hesitantly replied, "Love doesn't necessarily have to be expressed orally."

She wiped her tears. "You never wrote the words either. If you have any love at all for me, how would you express it?"

"That's easy, by not taking advantage of your virginity."

"Oh, is that your goal? Are you saving me for a rainy day?"

Mario was edgy, "I have respect for you which means that I have love for you." Santina realized that it was the first time she ever heard the word love from his lips. She tried to analyze the last statement and knew from his reputation that she was probably the only girl who didn't share casual sex with him. The possibility for marriage was weak but did exist. In a strange unexplainable way, he made some sense. Santina thought: He's going by way of Chicago to arrive in New York. "I appreciate what you're saying but I have a wierd feeling about your trip."

Mario raised his eyebrows, "Why, did you have a bad dream?"

"No, no, nothing like that."

"Then, what's bothering you?"

She hesitated momentarily, "I think you are involved with another woman and I think she is in Italy."

Mario retorted, "Young lady, you are crazy. I was only involved with the war. People think that I have an affair with every girl that I meet."

The waitress finally arrived with the swordfish and Mario tapped her on the leg. Santina missed the pass and asked, "Are you telling me that you never screwed any girls in Italy?"

"I never said that."

"Now we're getting somewhere, the truth isn't so bad after all, is it?"

"Santina, a soldier needs a break once in awhile."

"Yes, Mario, I knew that my gut feelings were right."

"Don't allow what I've just said bother you. There were prostitutes recruited by Special Services while I was on rest and recuperation and remember, I didn't know you then."

She pursued, "Regardless, we are now getting somewhere."

"Wrong, we're getting nowhere."

She persisted, "Did you ever screw a woman that was not a prostitute?"

Mario was surprised with the line of questioning, "Do you mean in Italy?"

"Yes, I am referring to your precious *Italia*."

He laughed, "No problem counselor, the answer is no."

She sat back, "I'm sorry, but I love you so."

He asked seriously, "You do believe me, don't you?"

"Yes, I do, but my inner self doesn't."

"Believe it. I never had an affair with an Italian except for prostitutes."

Pico

"You're so vulgar when you're angry."

"Only if I'm accused of something that I'm not guilty of."

They finished the dinner and had an after dinner drink. Mario took her ringless fingers and kissed her hand, "We will talk more when I get back"

As they rose to leave she fantasized: If the clams take effect, I will know if he really loves me.

32

NO CHANCE FOR ANGELINA'S HAND

Dante Pandozzi was visiting Pico, the home of his parents for the first time. It was the spring of 1947. He was born and raised in America and was a product of a prosperous family. His parents advised him that he should visit the Italian branch of his extended family which consisted of numerous aunts, uncles and cousins.

Dante was a dashing and handsome man of nineteen. He was a spoiled youngster who was catered to whatever pleased him. Young as he was, women were his focus of attention. His parents tried, to no avail, to arrange a suitable marriage to a woman who could keep him in line. Recently, Dante got a young girl pregnant and it cost the family a considerable settlement to rectify the situation.

More reckless than his contemporaries, he was in the habit of wrecking cars during various escapades. In fact, he even once smashed his father's speedboat while showing off to a girlfriend. She ended up in the hospital with a broken leg.

His parents decided that it was time their son vacationed in Italy. There, he couldn't wreck a car because none were available. Boats were safe because he would be up in the mountains.

Their primary purpose was for him to marry a proper girl. His relatives were alerted to help attain his parent's wishes. The boy was wild and the quiet mountain village was the ideal locale to set his life on a proper course.

The letter of the patriarch of the Pandozzis demanded, "Find a good wife for my wild son, I don't care what it costs."

Needless to say, the Pico family branch invited to that extent, every female prospect to countless dinners and get-togethers. The girls were told that Dante was a prized catch. He had looks, money and personality and they all tried to impress him and the Pandozzis.

Dante was courteous to all, but one night he informed the relatives that the only possible wife for him was an American. He rated the local girls as a bunch of green horns.

One uncle joked with another, "It's a good thing that he has that view, we should have less pregnant women among us. He has behaved himself."

One early morning he was enjoying a coffee with an uncle at a sidewalk cafe in the main square. He was daydreaming as his uncle asked him, "What are you thinking about?"

"I don't know, there is nothing to do here, I'm bored and homesick for America."

His uncle took a deep breath and wiped his brow with a napkin, "It's going to be a beautiful day. Enjoy the serenity. Later, perhaps you can visit Rome."

Dante paid little attention to his uncle because his thoughts were suddenly interrupted by the appearance of a beautiful young girl who entered the town square. He felt as if he were struck by a lightening bolt and fell immediately in love.

She walked gracefully as her hair was mussed by the slight wind. Dante was speechless as he observed her pausing to make idle talk with people who went out of their way to greet her.

The American turned to his uncle Georgio, "Who is that girl that resembles Venus?"

Georgio's faced brightened, "My nephew, that is the one and only Angelina."

Dante lit a cigarette and blew out a perfect ring of smoke, "My God, she is the most beautiful woman I've ever seen. Why haven't you told me about her?"

"Forget her, she is in love with an American soldier and no one else will ever do."

"You don't know me, Uncle, I can win over any girl that I set my mind on."

"Not this *donna*, she is not only the most beautiful but also the most dangerous."

"Uncle, you must be losing your senses. That girl is truly an angel. She smiles and everyone smiles back with warmth and respect."

No Chance For Angelina's Hand

"Yes, they all respect her and for very good reasons. Angelina is very special to all of us, a young lady who is a legend."

The young Dante looked into his uncle's eyes and proclaimed, "I'm going to marry that girl. Nothing can stop me."

Georgio looked skyward, "You have everything going for you but you have no chance with Angelina."

Dante smirked. He had never failed with the ladies and was certainly confident the beautiful peasant girl had little chance in resisting him. The playboy rose from his chair and saluted his uncle in a mocking gesture, "Just sit here and observe the style of an American who is in love."

Angelina carried a wicker basket of yellow flowers from the Ginestra mountain bushes and was delivering them to her church as a tribute to her patron saint. Dante approached her and made a sweeping bow with a display of chivalry. "You are lovelier than the beautiful flowers you carry so gracefully."

"Who are you and what do you want?" Angelina responded with a surprised smile.

He waved a hand with an authoritative motion, "I am Dante Pandozzi and I intend to court and ultimately marry you."

She laughed at him, "Am I supposed to be impressed by your ridiculous way of making conversation?"

"Yes, it's only the beginning. May I carry your basket and walk with you?"

Angelina continued her walk and occasionally acknowledged a pleasant salutation from passing friends. "Mr. Pandozzi, I would rather you not follow me," she said pleasantly but firmly.

Dante changed his tactics, "Allow me to start over again. You are even more beautiful that I originally described and your voice has a musical lilt that captivates one's imagination.

Angelina laughed, "I think you should go back and join your friend."

"You interest me and I would be pleased to be your friend. Now, may I walk with you?"

Angelina irritated now said, "Pandozzi or whatever your name is, you neither interest me nor do you impress me in any way whatsoever. At best, your efforts are a waste of your time."

She entered the rear door of the church as Dante stopped in his tracks. He made one last effort as she disappeared, "I assure you, my love will grow on you."

Pico

He returned to the sidewalk cafe as Georgio asked, "How did you make out?"

"Great, I think she will realize that I am the perfect man for her. I will make her the Italian princess of America."

Georgio laughed, "Even if you were successful in courting her, she would never leave this area. Forget Angelina, she is not for you."

"Uncle Georgio, you laugh a great deal, but we will see who laughs last."

Dante spent the day inquiring about the Micheletti family. He was saddened by the hard times but was awed by most of the information which further fueled his pursuit. Dante had a powerful ace-in-the-hole best described as money. In America, it bought whatever one desired. He rationalized: Money has the same power in Italy as in America. All he could think about was: Wait until my parents and friends set eyes on this angel.

The next morning, Dante dressed in his most fashionable clothes and set off to visit the Micheletti family. He hiked the small dirt road and observed the local peasants working their small farms. It appeared that every square foot of arable land was planted. In fact, only the rock-laden zones were left untouched. While Dante drew some attention because of his flashy attire, he didn't notice as his thoughts were on Angelina.

A young boy was walking towards him. Dante greeted him and asked to be directed to the Micheletti farm. The boy pointed in the proper direction which was now visible, "Jump over this wall, it's a short cut."

Dante thanked him and leaped over the small rock wall and inadvertently stepped on cow dung. He cursed in disgust and scuffed his shoe on the nearby grass.

The birds were singing the joyful lyrics of springtime. The mountain air was fresh from the dew of the overnight clouds. The sloping hills displayed a fertile lush green dotted with the bright yellow flowers from the Ginestra bush. The silence interrupted only by the sounds produced by nature. Dante was captivated by the freshness of the air, the beauty of the hills, the scent of the wild flowers and the tranquility. An experience of a lifetime. He imagined to himself: I must be in paradise. This is my lucky day. Nothing can possibly go wrong. The American stopped again. He took several deep breaths of fresh air and acknowledged his parents were correct in their assessment of their homeland.

Anna was sitting in a chair enjoying a cup of coffee. She too, was admiring the clear view and the Abruzzi mountains. Dante reached the upper landing near the stone farmhouse. Anna saw him but ignored his approach. He tipped his hat and announced, "It's a beautiful day."

She looked at him curiously, "Every day that one is alive is a beautiful day. Who you are and what you want?"

"My name is Dante Pandozzi from America. I've come for your permission to court your beautiful daughter, Angelina."

Anna scanned the young man from head to foot and asked with a hint of arrogance, "Who in the hell is Dante Pandozzi?"

"Haven't you heard of the Pandozzi family in Pico?"

"Yes, I know the family. Are you bragging or complaining?"

Dante stunned quickly replied, "The Pandozzis in America are very wealthy. In fact, we could probably buy the entire town and turn it into a golf course if we so desired."

Anna sipping her coffee nearly choked as she burst out laughing, "Mr. Dante Americo Pandozzi, do you think that wealth impresses me? Can you possibly imagine that the word 'golf' would scare me? What is a golf course, an American concentration camp? Furthermore, I have no respect for wealthy individuals who reek of arrogance."

Dante backed off to collect himself and as Anna stared at him. After a few moments, he said, "I'm sorry. I only wanted to convey that I have the resources to provide for your daughter."

"Be advised, we are not living during the era of Caesar Augustus. Matchmaking is a thing of the past. I have no control over who my daughter marries. Angelina has everything she needs and is a very happy lady."

Dante thought for a moment as Anna's warm smile returned. Encouraged, he asked, "Does she have a boyfriend?"

Anna continued to smile pleasantly but retorted, "It's none of your business."

"Signora, may I trouble you for a cup of coffee?"

Anna rose and entered the house. Dante watched and was impressed with her beauty and lovely, stately figure. He reasoned that living in the mountains and farm work contributed to their well being and beauty. Anna returned and handed him the coffee.

Pico

Dante was awkwardly silent and for one of the few times in his life was at a loss for words. He finally asked, "Do I have your consent to court Angelina?"

"You don't need my permission. She has a mind of her own, and a strong one, I warn you. If you want to court her, ask her on your own."

He broke into a broad smile, "Thank you, Signora, I will make her happy."

Anna surveyed him again from head to foot and laughed as if she were contesting with a child. Dante squirmed in his chair and nearly dropped his coffee, "Why do you laugh?"

"You are so confident. You do not have the qualities to impress Angelina."

"How can you make that evaluation, you don't even know me?"

"Yes, I do know you and the Pandozzis. You are all lazy and good-for-nothing."

"I won't argue that point but the American Pandozzis are different. Your daughter will be proud to be part of my family."

Anna spoke softly and sincerely, "Dante, Angelina will someday marry a special person who possesses class. She can never be attracted by a *cafone*."

Dante sat back and propped his feet up on a nearby stool. He smiled assuredly, "I was told on many occasions that I have that rare quality referred to as class."

She scanned him with scorn, "Look at you! Dressed in tailor-made clothes, nice haircut, clean shaven and ladened with wealth, but you lack class. You are a *cafone*."

"Me, a cafone, are you serious?"

"Mr. Dante, no matter how much you try, there is still a trace of manure on your fine shoes, no class."

He looked at his shoes and noticed a vestige of cow dung on the heel. They both laughed as Anna toyed with him.

She felt that he wasn't all that bad and somehow enjoyed his company. Anna looked at him and pointed towards the Abruzzi mountains, "There isn't a road that goes through those mountains. It's the same with Angelina. No one will ever enter her mind unless she allows it to happen. She endured the ravages of war while looking the devil in the eye with a challenge. Angelina, like the very angel that she is, came out the winner. The only man for her

No Chance For Angelina's Hand

is one with more courage than her. A person who had, likewise, challenged the devil and came out an even greater victor."

"Signora, perhaps, such a person does not exist."

"That person does indeed exist and she has met him."

Dante either didn't understand or didn't care as he asked shyly, "May I try to convince her of my intentions?"

"I won't stop you but don't feel rejected when she ignores you."

"Where do I find her?"

"She is working in the fields with her sister and brother-in-law." Anna pointed eastward, "You will find her there."

Dante arose from his chair and fixed his tie, "Thank you Signora, I'll see you later."

Angelina looked up from her planting chore and saw Dante approaching from a small wooded area. She remarked to Gabriella, "Look at that idiot, he can't even find the clear path."

Gabriella quizzed her with suspicion, "Who is he?"

Angelina returned to her planting and replied without looking any further at Dante, "Only another asino from America."

Horst offered, "Do you want me to get rid of him?"

"No, he is both amusing and harmless."

"What does he want?" Horst asked.

"I met him in town yesterday and he wants to marry me. Can you imagine the audacity of some people?"

Gabriella laughed, "What if that was Mario walking towards us?" Angelina's face tensed, wanting to vent her anger and disappointment over Mario by retorting bitterly. Instead, she resumed working.

"Don't forget about my dream, Angelina, Mario is coming."

"Your dreams are a farce. He is having too much of a good time in America. By the time that he wakes up, I'll be an old maid."

Her sister hugged her, "I had another dream and Mario was handing you a bouquet of Ginestra flowers. He will be here before the month of June."

Angelina wanted desperately to believe her sister's dreams, "Yes, June of what year?"

"This year. He came just in time for my baby's birth and that birthday is in a few weeks."

"Gabriella, don't make up fantasies. We haven't seen him in years. In fact, it's three years this month and not even a letter."

Pico

Horst smiled, "Don't give up. I don't believe in dreams but I say, Gabriella's do happen to come true."

Dante arrived and removed his hat and bowed to Angelina. "*Cara mia*, I am concerned that you are working too hard."

Angelina was not amused as she straightened up, "I'm not your *dear*. My toiling is none of your affair and what in the hell are you looking for?"

He ignored her outbursts and introduced himself to Horst and Gabriella. He wished them good luck for their child's future birth. Horst shook his hand and intentionally gripped it tightly in a vise like grip.

Dante's hand turned white. He smiled grimly as the German released his hand. Angelina and Gabriella laughed. Gabriella was proud of her husband. Their love grew stronger each day and it was probably the only positive matter that came out of the war for them.

Horst wiped his brow, "What brings you to our beautiful mountains?"

"I'll come straight to the point. Are you responsible for Angelina?"

"Yes, you may say that I am and certainly I'm concerned for her welfare. Why do you ask?"

"If indeed you have the authority, I want your permission to marry her."

Horst burst out laughing as Angelina mocked a swinging motion with her hoe towards Dante. Gabriella stepped between them as Angelina said in disgust, "Go back to America."

Horst intervened and put an arm around Dante as he led him away from the girls. He explained, "Every bachelor in the province wants to marry Angelina. The young lady is very capable of making her own decision. She will be the perfect wife for the most deserving man who can tame her."

Dante thought for a moment, "I could be that person. I have money, prestige, looks and personality."

Horst looked at him as if talking to a child, "I don't want to hurt your feelings but the best way I can explain the situation is that you and she are in different leagues."

"I realize that and know that, in due time, I can teach her to accept and conform to the other ways of life."

No Chance For Angelina's Hand

Horst looked in disbelief, "You have the assessment backwards. Angelina has already accomplished greatness. She will become a notable woman of her time and is, without a doubt, the most famous and most loved person in Pico. Angelina has special status and she earned it."

Dante tried to comprehend Horst's words and suddenly wasn't sure of himself. His confidence was ebbing, "Do you mean to imply that this girl here with a hoe in her hand is a goddess?"

"Very close to it. Wiser words are seldom chosen. She is beautiful, pure, strong, brave, thoughtful and every other attribute becomes her."

Dante replied with caution, "I have found the very person that I desire. I knew it from the first time I saw her."

"Congratulations, but you are doomed to failure."

"Why?"

"Angelina met her man during the war and her love for him grows stronger with each passing day."

"Who is he and where is he?"

"My friend, it is no concern of yours. My advice is that you direct your intentions elsewhere." Horst said it more as an order rather than advice.

Angelina and Gabriella began singing as they quit for the day. They were oblivious to Dante and ignored him as they headed back to the farmhouse. Dante was subdued and kept his distance from Angelina.

Anna had prepared dinner and much to Angelina's protest invited the American to join them. Dante did most of the talking as he related his life story which was impressive by American standards. He played high school baseball and football and explained that he was presently on vacation from college. He had a new expensive automobile and consequently had everything a typical young American man desired. Dante spoke passable Italian but at times combined it with English.

Dante had the floor as the others patiently listened. It was more a gesture of respect than of interest. Finally, Dante asked, "Tell me something about your experiences."

They were all quiet. A few moments, Anna chose to answer, "Nothing much, except that we experienced a recent war and encountered a great deal of hardship. Our education is adequate.

Pico

We have very little money which is not a factor for happiness. We do not possess a car but we had a jackass that died. We have a grove of olive trees and Horst is building a small olive oil factory. As for games, we enjoy an occasional play of bocce ball. We are humble, religious people who appreciate and enjoy the life that God is giving us."

"Everyone is healthy and as you can see, Gabriella is with child. The family is expanding," Horst added.

Dante was embarrassed, "I'm sorry but we Americans tend to brag."

Anna smiled, "It isn't necessary for you to apologize. You have a different lifestyle and consequently different values than we have."

Dante rose to his feet, "It's getting late, I want to thank you for an excellent dinner."

Anna joined him, "You are leaving so early?"

"I promised my family that I would be back home soon. They will be worried over my absence. May I call on you again?"

Angelina smirked in the background as Anna consented, "Of course, you are welcome anytime."

Dante left as rain began to fall slightly.

Angelina was disappointed with her mother, "Why did you agree for him to come back?"

Anna kidded, "The boy is harmless and wealthy."

Horst winked, "Who knows, he may present us with a jackass."

Angelina was furious, "I have no interest in that *asino*."

The German joked, "I think he is in love with Anna."

Gabriella showed support for her sister, "Don't worry, Mario will be here soon."

Angelina didn't reply and somberly, she walked outdoors to be alone. Her mother followed to join her. They sat in the misty rain absorbing the stillness of the dark night. Anna drew her daughter into her arms and comforted her, "My dear baby, and you still are my baby. Perhaps you should forget about Mario. We all love him very much but even though it isn't intentional, he has forgotten us."

Angelina looked at her mother with tears in her eyes, "I know that he will return but when?"

"Dearest, Americans are different from us. They have so many distractions in life. Look at your father and the rest of our American family. They hardly ever write and probably will never visit us."

No Chance For Angelina's Hand

"Mama, I don't care about them. I have never seen them. I'm sorry but out of sight, out of mind. Out of respect, I'll accept them when they visit me, but not looking forward to it."

Anna tried to console her, "Here, there are many things that remind you of Mario. It's only natural because you are constantly reminded of him. It's unfair to you because he isn't faced with the same conditions. Mario is in America and you have never been there except when I was carrying you during pregnancy. There is nothing in America that can be associated to you. Angelina was silent as her mother continued, "I do not advise that you take an interest in the asino who just left here, but perhaps some fine young man in the area. There are many to choose from."

"Mama, you're the only person besides Gabriella and Horst who understand my feelings. Mario is the only person in the world for me."

"I agree but you may have to wait forever."

Angelina countered, "So be it."

Anna gently smiled, "You're only sixteen years old and certainly have time on your side. Meanwhile, what are we going to do with this character, Dante?"

"Mama, I will take care of him tomorrow." It rained all night and Angelina welcomed the raindrops on the tile roof. She was getting weary of her heart beating for Mario.

The next morning produced cool fresh air that rolled off the mountains. Angelina drew deep breaths as she enjoyed a cup of hot coffee. Just as she suspected, Dante was approaching along the dirt lane. He didn't take the short cut over the wall and much to her surprise, he was wearing a simple attire. He wished her a good morning and walked by her towards the farmhouse.

Angelina stopped him and took his arm, "Come, Dante, let's take a walk and talk a little." She gave him a gentle smile and he was delighted.

The host pointed out the Ginestra bushes that were in full bloom. "They are my favorites. I love the yellow colors. Have you ever seen them before?"

"Yes, a few days ago when you delivered a basketful to the church."

Angelina brushed her hair, "The Ginestra has always been a significant reminder of important events of my life."

Pico

"Will you tell me about it?"

"I consider it the Easter flower. The bond with Jesus and the resurrection is symbolized by the flower. Every Easter since I can remember, I associate the flower with the holiday. I can recall picking the first buds and giving them to my mother. When my grandmother died of starvation during the war, we buried her and planted the Ginestra on her grave. It was all that we had." Angelina pointed to the area near a large fig tree where the deceased was laid to rest and continued, "It appears that everything around here happens in the springtime, both good and bad."

"What else happened?"

"Three years ago, the front lines of the war was concentrated in this area. They shifted back and forth until the Goumiers took control and committed atrocities. The most important occurrence of all was when Mario Calcagni saved this area from the Moroccans. The people were grateful but they had no better way to show their appreciation than to shower him with Ginestra flowers." She had a lovely smile but her eyes misted, "The Ginestra took on another role by honoring a great war hero."

"Who is this Mario Calcagni?"

She eyed him with amazement, "You have never heard of him? He is not only one of the biggest heroes of the war, but he is also a Picano with family roots here."

"I was only joking. Everyone has heard of him, how does he concern you?"

"That is why we are taking this stroll. I'm telling you Mario and I are going to be married someday."

Dante laughed, "You're only a child with rocks in your head. He is famous and has hundreds of girls chasing him around."

Her face took on an expression of scorn, "I love him and he loves me."

Dante tried to discourage her, "How can he possibly be in love with you when you were only a child of thirteen?"

"I saw the love projecting from his eyes as I handed him a bouquet when he departed."

"Angelina, listen to me. It would be impossible for him to come here and marry you. He is already settled in the lifestyle that he is comfortable with. Can you imagine Mario Calcagni with a hoe or shovel in his hand? Let's face reality, he would be more comfortable

No Chance For Angelina's Hand

with an automatic weapon in his hands and there are no conflicts on his horizon."

Angelina posed rigidly, "Mario will be here and I will be waiting."

Dante fired his best shot and realized that it had no effect. He knew he failed to impress the beautiful peasant girl. His confidence was destroyed but said, "You will wait as many Italian peasant women have and that, cara mia, is forever."

"Mr. Dante Pandozzi, we peasant girls are graced with the virtues of love and patience and those will prevail."

Dante tried one last time, "Do I have any chance with you?"

"No, you never had and you never will."

"Goodbye, Angelina."

She didn't reply. It wasn't necessary.

The Sunday Mass concluded as the bells rang out loud and clear. Anna met Anita on the church steps as Angelina, Gabriella and Horst walked ahead. Anna embraced her best friend and asked, "Where are Giovanni and the children?"

"They attended the early Mass and are visiting his mother. She is not doing well. I haven't seen you for a few days, where have you been keeping yourself?"

"I've been tied up. Haven't you heard of the American, Pandozzi?"

Anita laughed, "I heard he took some of your time. Let me guess, another Angelina admirer?"

"You guessed right, but he got the message and is gone for good."

The two matrons walked happily along and stopped a while to accept flowers from a little girl. The years have treated them well. After the war, they regained their health, beauty and vigor. Everything was going very well. Anita smelled flowers, "Ah, the fragrance is sweet." They stopped by the fountain in the square and Anita asked curiously, "Aren't you going to tell your friend about Dante Pandozzi?"

"There isn't much to relate. Even though he is spoiled because of his family's wealth he is not a bad sort." Anna thought awhile, "I'm a little concerned for Angelina."

"Why, she looks great and appears to be a happy young lady."

"You're right but how many eligible males can she turn away?"

Pico

"Anna, we all know the reasons for her decisions and patience. Try to have a positive outlook for her future."

Anna looked up to the heavens and pleaded, "Mario, Mario, where are you?"

Anita consoled her, "He made a contribution here that will never be forgotten."

"I'm with you. He was with us for only a few days and his exploits become more legendary each day."

"Why doesn't Angelina write him?"

"Are you kidding? I mentioned it a few times and she says that it isn't necessary. My sweet Angelina has more optimism than logic allows."

"Anna, please don't give up on Mario. He must have all sorts of diversions in America and he is probably biding his time. He will recognize the feeling and everything will be alright."

"Anita, did you know that he once told her that he loved her?"

"Yes, I know. It happened when Mario and Angelina rescued us from the cave."

Anna thought for a moment, "He actually didn't say the words."

"How could a child accomplish such a feat regarding love?"

Mario was vulnerable at the time. He thought Angelina was seriously hurt. She asked, 'Would you love me if I were five years older?'

"How did my nephew reply?"

He answered, "Yes."

"Anna, do you truly want my opinion?"

"You know that I've always valued your words."

"I married a Calcagni and to this day, I wouldn't trade him for anyone. Giovanni is a real man. I'll tell you one thing that I've experienced."

Anna raised her eyebrows in anticipation, "What do you tell me?"

"Once a Calcagni falls in love, there would be no other woman in his life."

"Well, what about Mario?"

"I'll tell you about him. He left here three years ago and associated with many girls in America but never got serious to marry any of them. He is like a soldier running through a minefield."

No Chance For Angelina's Hand

"That's what I don't understand. Why hasn't he found the right girl in America?"

"Anna, I saw that Calcagni look in his eyes when he spoke with Angelina during the rescue at the cave. He didn't comprehend what happened but his eyes gave him away." Anita wanted to take a more active part as a matchmaker, "Should we take matters into our own hands by writing him?"

"No, for the love of God, Angelina would never forgive us. I only pray for the day that he comes back but I have my doubts."

Anita took Anna's hand, "I know that America has a hold on him but we have one thing going for us."

"What's that, my friend?"

"Mario is a Calcagni and Angelina is on his mind. They were placed on this earth for each other. In their case, continents cannot keep them apart. Fate will take its course and ultimately prevail."

33

THE UNIFICATION OF LOVE

Mario got his travel affairs in order. He and Dean Jackson boarded an army transport plan and left for Italy. He knew that the strange feeling tugging him for the past three years was about to surface. Mario was now driving a jeep through the town of Cassino. Dean sat beside him in silence, he began to draft the first pages of the war efforts of Mario, the hero.

Cassino was an old community built by the ancient Romans. Then it was known as Casinum. Records reveal that prominent Romans vacationed in numerous spas in the surrounding foothills.

The entire town was nearly destroyed during the war. Rebuilding was now being accomplished at a fast pace and Cassino was on its way to recovery and consequently reestablish the area as a tourist attraction.

Mario pointed out the various locations that were his sphere of involvement. Dean was careful not to probe. His approach was not to deluge Mario with questions. He would rely soley on Mario volunteering information. The former soldier pointed to a lush green mountain area and offered sadly, "The weather was as formidable an enemy as the Germans. The trails were muddy and rocky. It was always cold, damp and windy, there was no respite. The edges of the trail were littered with decaying bodies of men and animals." In remembrance he shook his head in disbelief, "The stench was overpowering but fear of death from enemy mortar rounds took one's mind off the chaos. Hell, we were not even sure it was enemy fire or our own fire. The orders were always: move up and dig in."

The Unification of Love

Mario turned left and continued along Route 6. He was now acting as a docent. "This road was and is still referred to as the Via Casilina. It really is the main road to Rome. In fact, the Romans built the road about 400 B.C. It's probably regarded as one of the most important highways in all of history."

Looking up into the mountains, he continued, "The German observation posts had every square foot of passable terrain covered. I heard they were short of ammunition, but I never believed that bullshit. The artillery and motor barrages were constant, I still believe that some of it was from our own people. Can you imagine what those fuckers could have accomplished if they had the armaments that we had?" He turned to the writer and asked him, "Well, what do you think?"

"I guess the situation would be a stalemate."

"I don't think so. Those Moroccans would have still broken the line."

Mario glanced at the area of the Monte Cassino Monastery. He didn't want to discuss the Abbey, but offered one bit of information, "The Germans were never in the Abbey, the fuckers destroyed it for nothing."

"I always wondered why the Allies destroyed the monastery," Dean said.

Mario's hands tightened on the steering wheel. "It was stalemate, both sides needed a headline. Frankly, I believe some important changes of command were on the horizon. Why did the Allies invade Italy? From what I've read, the British were worried about the Russians and were concerned that Stalin would sign a separate peace with Hitler unless he received some relief with another war front."

Mario replied with a puzzled expression, "Italy is the art center of western civilization. They say that the entire country is a museum. American boys should never have been subjected to such a ridiculous situation. The New Zealanders were led by an idiot who didn't know his ass from his elbow. The poor Polish contingency was led down the primrose path. They were clobbered on Monte Cassino, suffering tremendous losses and got nothing out of it except for a cemetery on the other side of the mountain. The deal was already cut. The Russians would take control of Poland. The poor bastards thought they were making a contribution

Pico

to Poland. The sad part about it all was that they never conquered Monte Cassino, the Germans left! I'm getting ahead of myself and didn't answer your question. The second front that relieved the Russians should have occurred either along the western or southern coast of France, or both."

"That sure makes more sense, less mountains and a shorter distance to Berlin," Dean offered.

"True, Mr. Writer and the logistics would certainly have been an advantage. Keep in mind, the shortest distance between two points is a straight line. The straight line between London and Berlin was also a short distance."

Mario turned the jeep into the Liri Valley and was now expansive. Dean kept quiet and listened, enjoying the scenery. "Now take the British, I now hear that it was their idea to attack Italy! The United States was against it. Can you imagine the British setting policy? Hell, if we didn't come into the war, they would have been history. The French." Mario dropped the subject and sighed.

Dean knew that the French aspect, especially the Moroccan Goumiers was a taboo and he changed the subject. "It's truly amazing how the peasants use every bit of arable land."

"Yes, they are a resourceful people but the accomplishment was a necessity."

Mario came upon a bend in the road and pulled over to the side and stopped.

"Why are we stopping?"

For the first time since they met, the hero laughed and announced, "Piss call."

The writer returned the laughter, "I'm sure the Roman legions stopped for the same reason."

Mario added quickly, "So did the Americans."

They continued their drive and Mario was quiet. His mind went back into the past and was thinking about his parents when he told them he had to go back to Pico. His father questioned his intentions, "Why go back? The memories are not good for you." The old man had taken another sip of his homemade wine and waited for an answer. Mario remembered for a while before replying, "I don't know, Pop, there is something inside me that's bothering me. It's drawing me back there, I can't explain it." Mario had moved his

The Unification of Love

empty glass towards his father who refilled it. *"Mio figlio,* I remember when you didn't like wine."

"Pop, good wine is one of the pleasures of life. I learned in Italy that it has medicinal value. Pop, believe me, the feeling I have is not bad. I can't describe it. It's a great feeling but I can't figure it out."

The Abruzzi mountains loomed in the distant east while the Aurunci Mountains lay peacefully in the west. Dean interrupted Mario's train of thought, "I can't imagine how soldiers could have fought in such a mountainous country."

Mario replied, "Neither can I. In fact, if you examine a map of Italy, you'd wonder how they survive. Italy is a strong Catholic country with more churches than you can imagine. Consequently, they follow their religion and have many children. There is not enough space for an expanding population, so many immigrate to the United States, Argentina, Venezuela and on and on. Hell, I've got relatives in Argentina and Brazil that I don't know about! Southern Italy has produced illiterate people who immigrated to the United States, but generations have actually produced doctors, lawyers, political leaders and yes, even mafia dons."

They both laughed, but soon after, Mario went back to memories of conversation with his father.

"Pop, why don't you and Mom come back to Pico with me?"

Rocco shrugged, "If my mother laid eyes on me, she will surely die from a heart attack. Don't forget, I left Italy when I was fifteen years old and never returned. Besides, after what she's been through...," his voice faded with sadness.

Mario placed his hands over his fathers, they both had tears in their eyes, "I understand, Pop, but someday when conditions are right, you must go back."

When Rocco learned about the brutality inflicted upon his mother, he stopped attending Mass at church.

They drove by Roccasecca and Mario smiled faintly, "I can recall that name Roccasecca because whenever I planted a tree back home, my mother would always say it leaned towards Roccasecca. Hell, I couldn't plant anything straight. She loved poking fun at that town."

Dean offered, "Did you know the German area commander, Von Senger, had his headquarters located in Roccasecca?"

Pico

"Yeah, I heard that he wasn't a bad guy. I read somewhere that he was a Catholic and was a lay member of the Benedictine Order."

"Isn't that the same group of monks that ran the Monte Cassino Monastery?" Dean asked.

"Yes, it is. A bit ironic, isn't it." Dean nodded in agreement.

They soon drove through Arche as Mario said, "My home town has a number of people who immigrated from this area. A big family with a long name. They own a grocery store, a bakery, carpenter shop and many others and work hard. I always admired the way they stuck together. Arche has also produced a mafia Don that is highly respected," Mario laughed, "by his peers, that is."

"I know it's too late, but why didn't we take route 82 through Itri?" Dean asked.

"We can drive that mountain road at a later time. I'll trace my exact war route." Mario was in a rush to get into the Pico area. They turned left at Arche and headed towards San Giovanni.

"I also know people who came from this beautiful town. My father had a compare that was a San Giovannese. The guy was a nut. He had a hunting dog that he treated better than his wife. Vito would purchase expensive hamburger for the dog and inferior meat for the family. On the first day of the hunting season, Vito fired at a rabbit and the dog ran off. He never found the dog which, by the way, was afraid of gunshots. He blamed the wife, accused her of wishing him bad luck with the dog. My father always said that his compare Vito was touched in the head."

Mario didn't say much about the little hamlet between Arche and San Giovanni called Ceprano. He knew very little about it except it had something to do with Venus and mythology. They crossed a little bridge before entering their final destination: Pico.

To Mario, it seemed he was seeing Pico with the eyes of an ancient Roman. Ah, this beautiful hamlet of intriguing history, of gentle people, and regal, quiet mountains. Mario drove slowly, admiring the serene surroundings. He was deeply engrossed and quiet.

Dean broke the concentration and asked, "Mario, can I ask you a question?"

"Sure, ask."

"I know you have close relatives here, but I get the feeling that you are here for some other reason."

The Unification of Love

Mario was tentative and did not reply until he drove completely out of town, "You know, you're right, but I don't know myself." Without realizing it, he pulled to the left of the road and stopped. The area was slightly north of town and familiar. He realized he had been on this very location before. Mario got out of the jeep and asked the writer to follow him. He appeared to be anxious and yet elated. Dean thought: This man is on a mission.

They hiked up a rocky section of a large hill that looked like a mountain. Dean was an alien to this type of terrain and had a difficult time keeping up.

Mario knew the way, but appeared he was being led by some unknown source. There was no path or trail but he knew exactly what direction to take. Dean was tugging on the brush to help himself along. Mario waited for the writer to catch up. He then took his hand and helped him the rest of the way.

They soon arrived on a reasonably flat acreage and came upon a well-constructed house and with a large barn. Dean knew by the look in his eyes, that Mario had been here before. He was exuberant and appeared about to sing a joyful song.

It was now late afternoon and Mario became serious. The sun was between mountains on the western horizon. Mario placed a hand on Dean's elbow and faced him north. He pointed menacingly with his index finger, "They came from that direction." Mario became enraged, "The fucking Goumiers outflanked the German line and surprised Pico from the rear. They crossed exactly where we are standing. The sons of bitches went by that house and headed down that trail towards Pico. They resembled locust in a wheat field, only, the wheat was the people of this town." His face was full of scorn, "It was unexpected. They were great mountain fighters but they committed atrocities." Dean had heard stories of how Mario reacted back in 1944, but he chose not to question him about them now.

"They brutally assaulted and raped every living thing they got their fucking hands on. When they were done with people they went after the animals! Dean was stunned. Mario continued his rant, "It figured because they were beasts themselves." He became subdued as tears came into his eyes, "Them fucking bastards raped an eighty year old woman six times." He broke down and cried, "She was my grandmother."

Pico

Dean wanted to comfort Mario by placing an arm around him but decided against it. Mario regained his composure, "They left her for dead but the old lady is still alive!"

Dean thought: The Congressional Medal of Honor recipient is human after all but under those circumstances, anyone would break down. This is probably what has drawn this man back to Pico, the fact that his grandmother is still alive. I won't ask him. Sure, I saw a soft side of him but he is one tough son of a bitch. Hell of a man.

Off in the background, they heard a man singing as he was herding his few sheep home from a grazing area. The sun was setting as Mario was looking towards the farmhouse. They had previously noticed a young girl picking yellow flowers on a far-off hill. They hadn't given it much attention, but she appeared to have noticed them and began descending in their direction.

As far as Mario could make out, she was probably eighteen years old and possessed a beauty that resembled the Madonna. Her hair was auburn but appeared to display a reddish tint in the afternoon sun. Mario saw her bright hazel eyes and she was smiling. She wore a yellow blouse with a kelly green skirt that fluttered in mountain breeze. The closer she came, the broader the beautiful smile. Her confident stride and beaming face revealed she knew the American hero.

Dean gazed at Mario who appeared to be dumbfounded. Mario returned the smile to this familiar angel and recognized her but couldn't place her. He wasn't embarrassed and he was calm. Their eyes never left each other. It was as if time was suspended for this blissful meeting. The writer was witnessing an event that had special blessing from heaven. Neither spoke, it wasn't necessary. Their presence alone was enough. The donna's eyes spoke of love that has withstood time and distance. And Mario felt the stirrings of the mysterious feeling that he had trouble naming for the last three years. Ridiculous, but could he be in the throes of a first love?

For the second time during the last hour, Mario revealed a nature that he is indeed human. The hero has fallen in love. Words would not come to either of them. They were both ecstatic.

He tried to call her by name but it eluded him. She wanted to scream out his name, but held herself back with her triumphant smile. Tentatively, he spoke softly, "Angelina?" Amazingly and as if on cue, she cried out, "Mario!"

The Unification of Love

The words both came out at the same time. The world stopped moving for an instant as the two embraced. They spun around as one.

Mario wouldn't let her go, he kept repeating, "Angelina, Angelina, my angel."

He finally let her down and looked into her eyes, this girl who defied the Moroccan Goumiers.

Tears came into her eyes, "They never touched me."

Mario hugged her and again feasted his eyes on the brave girl. They walked towards the farmhouse as Dean followed. Both he and Mario now knew exactly what drew Mario back to Pico.

EPILOGUE

It is imperative to remember that in the time frame in which this novel was set, Italy was no longer one of the enemies. Specifically, Italy joined the Allies in September of 1943.

Undeniably, Mussolini and his small segment of fanatic Fascists had captured the imagination of the Italians with vision of national and global glory since the 1930's. Thus, the Italians were led, en masse, into the unholy, infamous triumvirate called the Axis. But while the Italians were throwing their stakes in with the Germans and the Japanese, millions of Italian-Americans in the United States agonized for their *paisanos* and *Italia*. As early as the 1930's, despite the rise of Fascism, Italy did not consider the United States as an enemy. Conversely, and in fact, even as Americans entered World War II, they were reluctant to invade Italy. Ultimately, with pressure from the Allies, largely from Prime Minister Winston Churchill, the United States agreed.

The Prime Minister was concerned with relieving the pressure the Russians were facing on their front. He certainly deserves merit in assisting the Russians, but why didn't he utilize the plan to invade Normandy ahead of schedule? France, the most direct route in ending hostilities is located several miles across the English Channel. Why did Mr. Churchill insist on reaching Berlin through North Africa and Italy? Can one consider or even imagine defeating a formidable enemy in the mountainous terrain of Italy that, in all of history, was only conquered once from the south!

The Americans were strongly against the invasion of Italy. There were many justifiable reasons for their initial firm stand. They reasoned primarily that the shortest direction to Berlin was through landings along the mainland coast of the English Channel which provided all the advantages required for a swift conclusion to ending the war. (Note: A more rapid conclusion would have saved many lives, including many who perished during the Holocaust.)

The Americans also wondered: Why destroy the museum of western civilization? Every church, nearly every building is a work of art of which some are over two thousand years old.

Why attack a peninsula of high mountains and waste time as well as subjecting American lives to uncertain conditions? German soldiers who had an easier time than the American G.I.'s reported the conditions to be worse than the Russian front!

In reality, once General Bernard Montgomery faced the actual conditions in southern Italy, he correctly rationalized to leave his Eighth Army and bailed out. He knew there was no glory to be attained in trying to scale one mountain after another.

The Americans were aware of the many Italian-American families who were concerned for their relatives in far away Italy. How Prime Minister Churchill was able to convince the United States to invade Italy is a mystery for the ages. Of even greater curiosity, why did he ever pursue such an undertaking?

During September of 1943, Italy withdrew from the Axis and joined the Allies.

Shortly, General Mark Clark, the Commander of Fifth Army, armed three hundred thousand Italian troops and accepted them under his command. In addition, at least one Italian division was facing the German enemy on the Monte Cassino front. They were looking forward to liberating Italians on the other side of the Gustav Line.

The point stressed is that the Italians were anticipating liberation from their friendly American allies.

They survived the constant bombing and starvation. The hope that the Americans were coming was the only food for digestion. They prayed and wandered aimlessly waiting for their deliverance. As liberation appeared on the horizon, life in peace was becoming a reality.

Unknown to the Italians the immediate future would result in one of the greatest inhumane acts of cruelty and brutality ever inflicted on mankind. History will confirm the fact that the decision to invade Italy was, at best, a very bad undertaking. In reality, when the war ended, the Germans still maintained a strong defensive position in northern Italy.

A tremendous amount of credit must be given to the American soldiers who had to overcome one adversity after another in driving the enemy northward. The odds of success were nearly impossible. The decision makers placed them in one precarious position after another but they prevailed.

Up to the fourth battle of Monte Cassino, the Allies were probably losing the war! The British Eighth Army and the American Fifth Army hardly ever agreed on anything. General Mark Clark, the Commander of Fifth Army, had the power of the most industrialized nation supporting him. Yet, he was making lame, questionable decisions. He nearly sacrificed an entire National Guard Division from Texas near the Rapido River that later, initiated an investigation from the state of Texas. General Clark gave the final authorization to destroy the historical Abbey at Monte Cassino which served absolutely no purpose in the penetration of the Gustav Line. He failed to analyze the ramifications of mountain fighting. The planning and utilization of air cover, to assist and protect the infantry, was non-existant. The excuse was that the enemy was too close and the Allies would suffer casualties from friendly fire.

Try conveying or convincing that information to General George Patton.

There is no doubt that the Commanding General would soon be replaced.

The United States had a multitude of competent generals who were chomping at the bit. The greatest of them all, General Patton was sitting on the sidelines. Needless to say, the American commander was under extreme pressure. He turned to his ace in the hole, General Alphonse Juin of the French Expeditionary Corps. Here was a leader with a plan. In fact, he was the only influential commander that analyzed and produced the equation to break the stalemate on the Gustav Line. General Juin was highly regarded and consequently favored by General Clark. The French military leader was perhaps the superior general of the entire Italian campaign. He shall, however, always bear the stain inflicted by his Goumiers. Ironically, he would have accomplished the same fame without the authorization of atrocities. The man was a military genius.

The French Expeditionary Corps consisted mostly of North Africans. Many were Moroccans, most were of the mountain Goumiers. While the commanders of Fifth and Eighth Armies were amassing tanks, trucks and jeeps for mountain fighting, they were conveniently gathering every mule and jackass in sight. The motorized elements of war were creating a quagmire on every road and at every intersection. Tanks, trucks and jeeps were useless in the areas that counted: mainly the paths and trails utilized by pack animals. General Juin fortified his positions but even more so, was strategically located to mount an offensive over the most difficult of terrains.

The Germans thought opposing these impregnable positions would be easy and subsequently decided to defend the area with a minimum cast of soldiers. General Juin had an excellent battle plan and he was prepared for victory. To further assure and consequently solidify his chances to break the Gustav Line, he proclaimed what the Italian peasants now refer to as The White Paper Act.

General Juin probably never titled the ill-fated period. It was an edict issued by him that officially was to last for only fifty hours. He promised his Goumiers that, once the breakout occurred, they were authorized to plunder and pillage. Needless to say, the results of these criminal acts led to beatings, murder and disease. Fifty hours lasted for a few weeks! The Goumiers offered lame efforts against the Germans. They turned their efforts against unarmed Italian civilians.

General Mark Clark was informed during the early stages of the crimes against humanity. He was well aware of their reputation and once had Goumier women imported from North Africa to satisfy their sexual lusts. The Goumiers had a history of raping residents in occupied areas. The Fifth Army Commander reasoned that he solved the problem but failed to consider the fact that they were engrossed with caucasian women. General Clark issued a strong complaint to General Juin that the atrocities must immediately cease.

The French Corps leader instructed his officers to stop the carnage. He added that they were to show no mercy against the crazy Goumiers. He ordered that they were to shoot anyone who was caught committing barbaric acts and violence. It appears the orders were not strongly enforced because the fifty hour period lasted three hundred hours.

It is a known fact that the Moroccans didn't care for the French. They were only loyal to their leader. Isn't it possible that Juin could have stopped the massacre if he so desired? Hell was endured for nearly two weeks. The Goumiers were inhuman. It didn't take much urging to guide them into whatever direction a commander led. The barbaric incidents that occurred are too numerous to mention.

One survivor described the barbarians as wandering demons with long knives by their sides. They searched for the women with white skin, considering them irresistible and attractive. Without any warning, the demons violated them brutally without consideration for health or age. The unfortunate victims who resisted were savagely beaten.

Another peasant noted, "They moved over the rocky terrain with the ease of a mountain goat. The barbarians rushed into homes and raped the inhabitants. When they were through, they took what few belongings were available. It was horrible, they terrorized everyone in sight. The women ran and searched for hiding places to escape their wild lust. The Goumiers were soldiers of fortune who caught their prey without mercy."

In one small village, a mother offered herself in order to protect her small children. They raped her and then proceeded to violate the children.

Another survivor tried to explain before he broke down into tears, "There are no words to describe the scenes of violence and abuse that resulted in death."

An old man hiding in an attic described a scene, "The screams were horrendous. The beasts grabbed everyone in sight. Some women who resisted were hit in the face with a rifle butt. Others were slashed with their dreaded twenty inch long knives. As they violated their victims, they grunted like pigs."

One farmhouse contained an old man and his aging wife. They were successful in delaying them while the rest of the family escaped. The barbarians were angered to the point where they raped and brutally beat the old couple.

A peasant was horrified by the fact that the Italians were facing two wars: the German and the Goumier fronts.

There were many incidents when children witnessed the rape and beatings of their parents! In some villages, not a single woman evaded the violence inflicted by the barbaric hordes. A grandmother of eighty was raped six times! The men who dared to defend their women were beaten and subjected to the same rape ordeals. Whoever refused to satisfy their lust was savagely bloodied beyond recognition. Many were tortured and underwent beastial sex.

Some victims became mentally deranged and never recovered. Hundreds perished from venereal disease contacted from the rapes.

The Goumiers had no regard for religion. The barbarians raped nuns and priests. When people were not available, they turned their wrath and desires on animals!

The Moroccan Goumiers left a trail of blood but of a greater tragedy. Most were the terrible memories they etched into the minds of the survivors who are ashamed of the atrocities.

They want to forget and will not come forth to reveal their ordeals. It is a chapter of history they wish could be torn.

One certainly cannot vindicate General Juin. He had four divisions under his control and the Goumiers numbered approximately three thousand. Juin could easily have brought them under control because the Goumiers were no longer interested in doing battle with the enemy. It is assumed by the author that the General had no intention to rectify the situation. This is another unanswered question associated with the Italian war campaign.

In later years, during a news conference, General Juin was asked why he showed no compassion or no mention of the atrocities in his memoirs. He did not answer. In the final analysis, however, the blame lies on the commander of the Fifth Army.

The United States, has a distinct concept regarding military responsibilities and accountabilities.

The United States Navy has a saying that, for all intent and purposes the Captain of a ship is responsible for whatever occurs on his watch, good or bad. Likewise, the Commander of an Army is directly accountable for decisions made in his theater of operation. He is the sole authority and consequently the recipient of military accolades for successful campaigns. The ultimate reward for victory is a favorable position in history.

Conversely, he must shoulder the blame for ill-fated end resulting from directives issued by his command.

All generals want to be victorious and justifiably so. In the event they don't win, they are lost in oblivion. Some generals will go to extremes to attain that place in history.

As previously noted, General Clark was getting nowhere fast. He desperately needed a miraculous move to re-establish himself. The Normandy Invasion was several weeks away.

Once the landings occurred, there would no longer be newspaper headlines about the Italian campaign. He rationalized that the capture of Rome would give him the distinction of being only the second Commander in all of history to do so from the treacherous mountains from a southern approach. History will reveal that the general had an obsession with the successful capture of Rome. The problem he was faced with was to make it all happen before the Normandy invasion. Time was certainly of the essence and he pulled all stops, including the advice and antics of his best subordinate: General Alphonse Juin.

In order to attain his goal, he required the efforts of the French Expeditionary Corps to protect the flanks of his American divisions. He also needed General Juin to give the British a hard time as a delaying tactic to prevent them from liberating Rome. In reality, the race was on with Mark Clark's forces driving up the coast while Juin and his French forces provided the diversion.

The overall situation can best be outlined as follows:

1) General Mark Clark, the Fifth Army Commander, needs to redeem himself.
2) The liberation of Rome is the plum that could achieve the above.
3) The problem is time. Normandy Invasion is on the horizon. The British Eighth Army could beat him there.
4) The solution lies with the French Expeditionary Corps. Allow the effective General Juin some slack and the chances of success are close to being assured.
5) During the early stages of the break of the Gustav Line, news of the atrocities reached the Commander's Headquarters. He informed General Juin to refrain from the atrocities. He in turn, informed his subordinates. There was a great deal of wrist slapping and very little of "rocking the boat."
6) General Mark Clark attained his goal of liberating Rome and, in conjunction with the atrocities, was instrumental in allowing the German forces to escape northward.
7) His place in history is accompanied with asterisks.

In summation, the Captain of his ship looked the other way in order to satisfy his obsession with the liberation of Rome. The direct consequence is perhaps the most savage, criminal and humiliating acts ever inflicted on the human race.

AFTERMATH

Pico is special to this author. My roots are there, my parents and grandparents are from Pico. As far as I can trace the family tree, so were their parents.

My grandmother endured the hardships of the war and witnessed the atrocities. I still have a aunt living there who lived through the experiences. There are numerous first cousins of who reside in Pico and in Rome. At the time, some were children and others not yet born. I have still never met my Italian cousins but knew who they were. I have never been to Italy and they have not visited the United States.

Halfway into my novel, I decided to contact my cousins.

I set a translator into my computer and wrote to six cousins requesting information regarding the Italian view of the terrible events. Two months passed and I didn't receive any correspondence. I was saddened and expressed my disappointment to my wife.

She replied, "I don't understand it. I knew their mothers and fathers to be reliable people."

I thought for awhile and replied, "I have no right to bother people whom I've never met."

She countered, "Don't feel that way, they are your first cousins."

A few weeks passed and I received a package of two books from my cousin Georgio from Italy. I was overjoyed and related to all my friends that I have a cousin. My wife Elena was happy because this gentleman's father was her hero during the most crucial point of the local battle. My uncle Cammile had returned from the Russian front and saved many lives by leading refugees to the safety of caves. He was a legend and much of the Calcagni characters where based on his life. My wife had tears of joy in knowing that of all my uncles and aunts, it was the son of Camille who replied.

I did my best to translate the books and realized why it is so difficult for them to take interest with a part of history: this particular history is one they prefer to forget.

There was one chapter in one of the books that substantiated all that was related to me by my Elena. The humiliation endured by the victims was so gross, so grotesque that few will surface with information.

One evening, my wife and I were sitting on a bench near our wine cellar looking eastward towards the San Diego mountains. She sighed, "They are not as high but they resemble the Abruzzi Mountains." I starred silently as she pointed out the comparison. Then she looked at me and asked, "What are you thinking about?"

I hesitated briefly and then replied, "I don't think I should go any further with this book. I'm considering shelving it."

"Why?"

"Because I'm opening a chapter in history that Italians would rather close forever. I believe I could never interview a person who was a victim of the savagery nor would that person willingly disclose the horrors!"

Elena replied firmly, "I was there and saw everything that was written in your book about the atrocities. I'm not afraid to talk about it."

I looked into her hazel eyes and smiled with approval, "Okay, Angelina, let's go with it!"

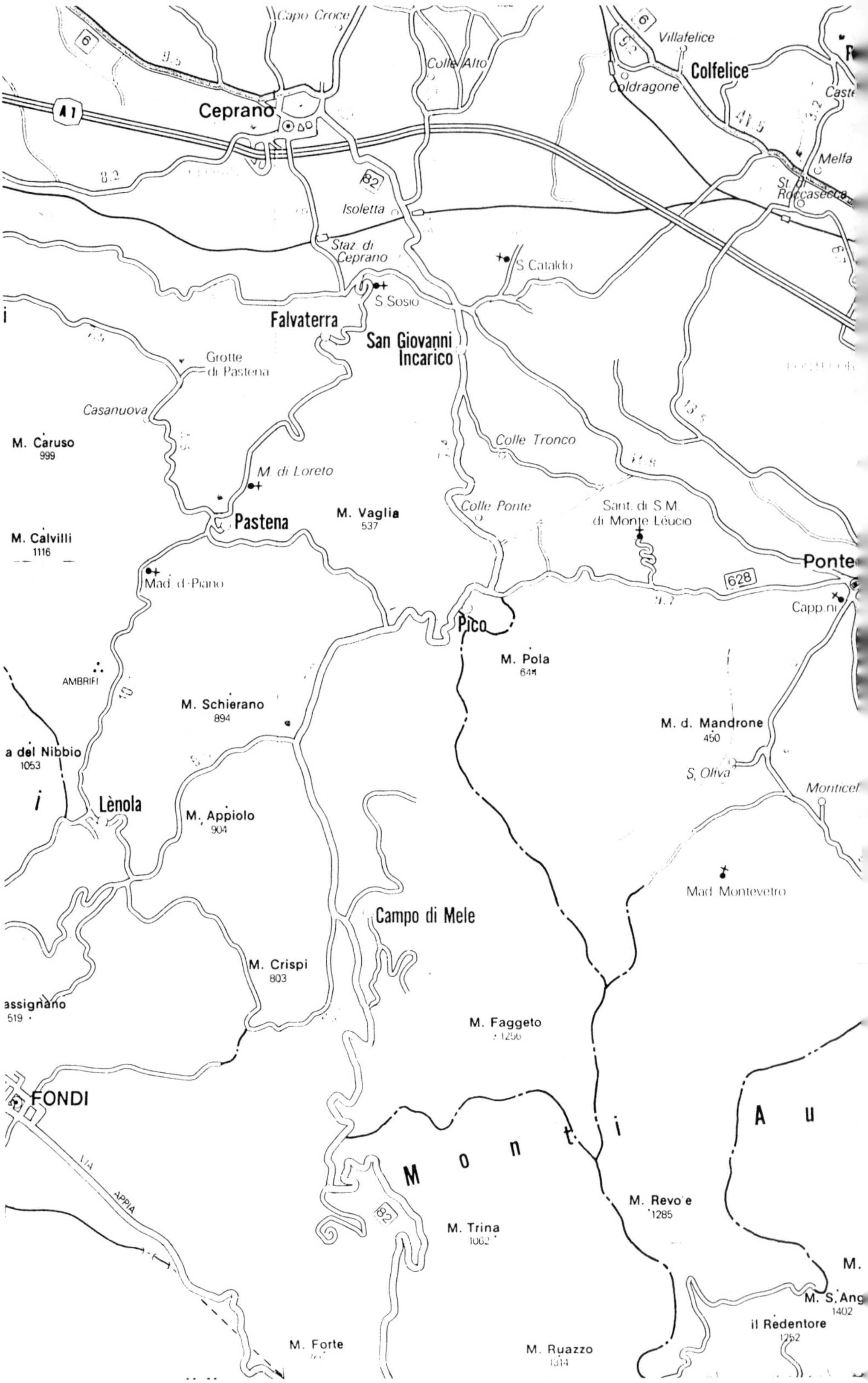
Capo Croce
Colle Alto
Villafelice
Colfelice
Coldragone
Ceprano
A1
6
82
Isoletta
Staz. di Ceprano
S. Cataldo
S. Sosio
Falvaterra
San Giovanni Incarico
Melfa
St. di Roccasecca
Grotte di Pastena
Casanuova
M. Caruso
999
M. di Loreto
Colle Tronco
Colle Ponte
Sant. di S.M. di Monte Léucio
M. Vaglia
537
Pastena
M. Calvilli
1116
Mad. d. Piano
628
Ponte
Capp.ni
Pico
M. Pola
644
AMBRIFI
M. Schierano
894
M. d. Mandrone
450
a del Nibbio
1053
S. Oliva
Lènola
M. Appiolo
904
Monticel
Mad. Montevetro
Campo di Mele
M. Crispi
803
assignano
519
M. Faggeto
1256
FONDI
VIA APPIA
Monti
Au
M. Revole
1285
M. Trina
1062
M. S. Ang
1402
il Redentore
1252
M. Forte
M. Ruazzo
1314